Name _____

Dancing Digits

Use the numbers **1**, **3**, and **9** to write a three-digit number on each line.

1. What is the largest number that can be made? _____

2. What is the smallest number that can be made? _____

3. What is the largest number that can be made with a **3** in the hundreds place? _____

4. What is the smallest number that can be made with a **3** in the ones place? _____

5. What number is greater than **193** but less than **391**? _____

Use the numbers **2**, **4**, **7**, and **8** to write a four-digit number on each line.

6. What is the largest number that can be made? _____

7. What is the smallest number that can be made? _____

8. What is the largest number that can be made with a **4** in the thousands place? _____

9. What is the smallest number that can be made with a **7** in the hundreds place? _____

10. What number is greater than **7284** but less than **7482**? _____

Brainwork! There are **24** different four-digit number combinations that can be made using the numbers **1**, **5**, **6**, and **9**. Write the six combinations that begin with a **9**.

© Frank Schaffer Publications, Inc.   FS-32022 Math Activities

Name_____

Skill: Patterns

Shape Patterns

Look for patterns. Cut and paste the shapes that come next.

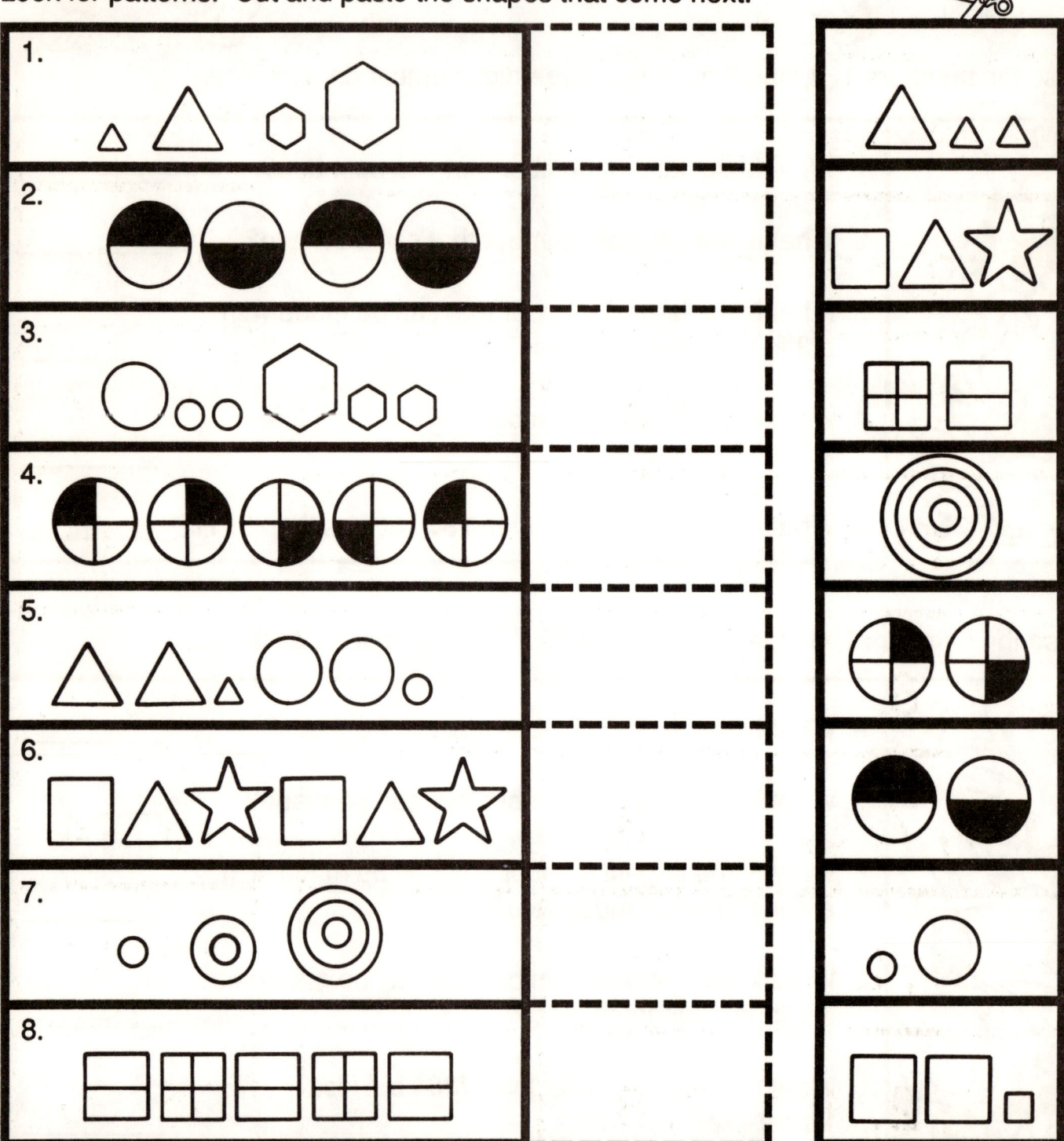

1.
2.
3.
4.
5.
6.
7.
8.

Brainwork! Draw a pattern using shapes. Ask a friend to look for the pattern and draw the shape that comes next.

© Frank Schaffer Publications, Inc.

2

FS-32022 Math Activities

Name_____

Skill: Patterns

Number Patterns

Look for number patterns. Cut and paste the numbers that come next.

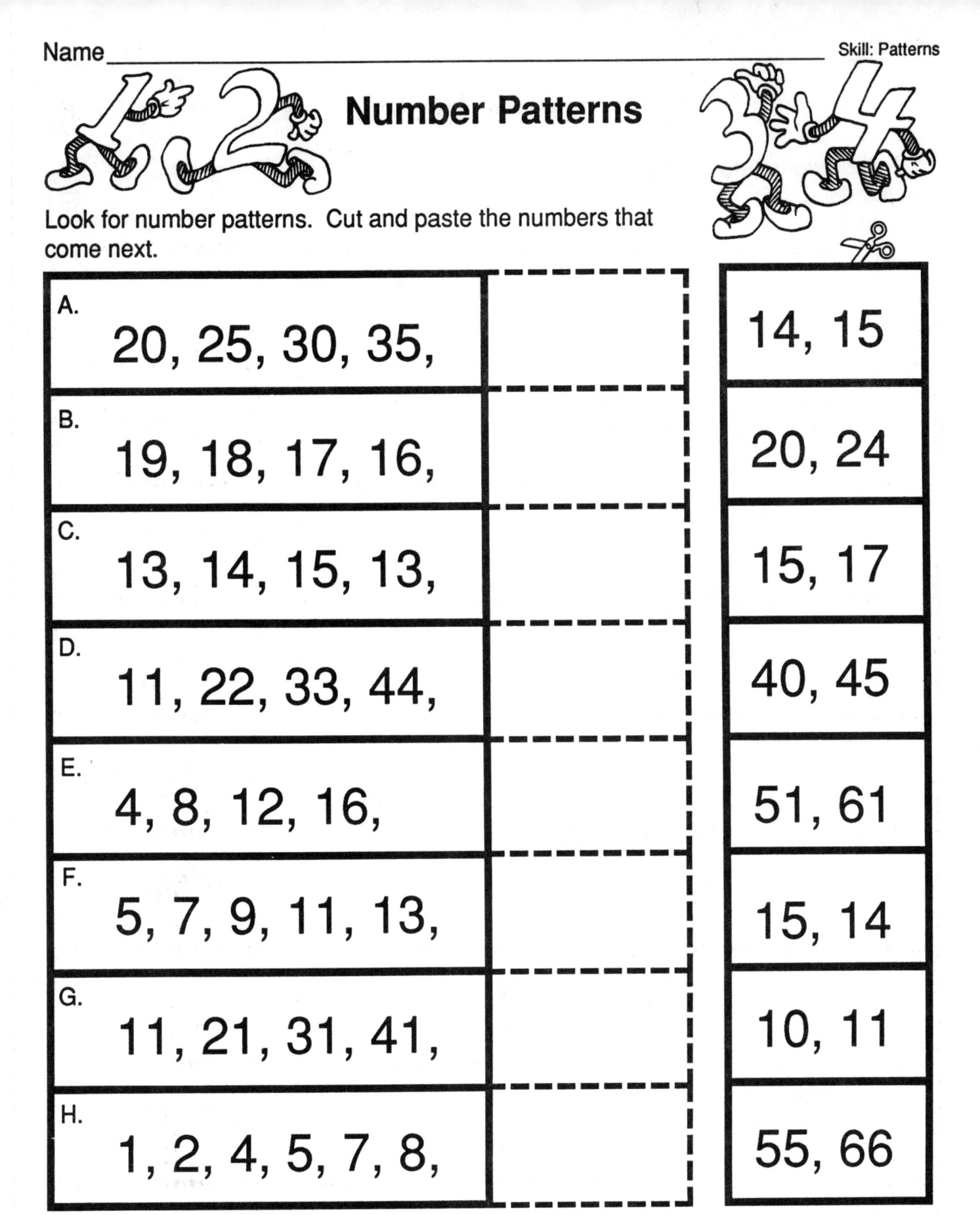

A. 20, 25, 30, 35,	14, 15
B. 19, 18, 17, 16,	20, 24
C. 13, 14, 15, 13,	15, 17
D. 11, 22, 33, 44,	40, 45
E. 4, 8, 12, 16,	51, 61
F. 5, 7, 9, 11, 13,	15, 14
G. 11, 21, 31, 41,	10, 11
H. 1, 2, 4, 5, 7, 8,	55, 66

Brainwork! Create a number pattern for a friend to solve.

© Frank Schaffer Publications, Inc.   FS-32022 Math Activities

Name _____

Skill: Addition key words,
Addition facts 11-18

Game Time

These key words are often clues that tell you to add. Use them to help you solve addition problems.

**in all
altogether
total**

Read each problem below. Underline the key words. Then write the number sentence and the answer.

1. Four friends were playing a game. Mai counted 5 question cards. Zach counted 7. How many question cards did they count in all?

   _____
   _____ question cards

2. There are 6 bonus cards to place near "Start" and 8 bonus cards to place near the dragon. How many bonus cards are there in all?

   _____
   _____ bonus cards

3. Mai's first card let her move ahead 3 spaces. Her second card let her move ahead 9. What was the total number of spaces Mai moved?

   _____
   _____ spaces

4. Ricky's first card let him skip ahead 4 spaces. His second card let him move ahead 8. How many spaces did Ricky move altogether?

   _____
   _____ spaces

5. Zach could win if he could get an 8 and a 7 on his next two cards. What is the total number of points Zach needs to win?

   _____
   _____ points

6. Lydia won the game. She moved ahead 9 and then 7 spaces. How many spaces did she move altogether in her last two turns?

   _____
   _____ spaces

Brainwork! Write three addition word problems about a game. Use different key words in each. Have a friend solve them.

© Frank Schaffer Publications, Inc.

FS-32022 Math Activities

Name _____

Skill: Key Words,
Subtraction facts 11–18

Bicycle Facts

These key words are often clues to subtract. Use them to help you solve subtraction problems.

how many fewer
how many more
how many left
how much less

Underline the key words. Write and solve the number sentence. Write the answer.

1. 16 third graders rode their bikes to school on Monday. Only 7 rode their bikes to school on Friday. How many more third graders rode their bikes to school on Monday?

   _____

   _____ more third graders

2. 8 children in Mr. Lopez's class have ten-speed bikes. 14 have coaster bikes. How many fewer ten-speed bikes than coaster bikes are there?

   _____

   _____ fewer ten-speed bikes

3. Jill can ride her bike to school in 9 minutes. It takes Kara 13 minutes to ride her bike to school. How many more minutes does it take Kara to ride her bike to school?

   _____

   _____ minutes more

4. It takes Ryan 15 minutes to tighten his brakes. Marco can tighten his brakes in 7 minutes. How much less time does it take Marco to tighten his brakes?

   _____

   _____ minutes less

5. One weekend Su-Lin and Ivy rode their bikes 5 kilometers. The next weekend they rode 12 kilometers. How many fewer kilometers did they ride on the first weekend?

   _____

   _____ fewer kilometers

6. In June the hardware store had 18 bicycles for sale. They sold 9 of them. How many bicyles were left to sell?

   _____

   _____ bicycles left

Brainwork! Write a subtraction word problem about bicycles. Use subtraction key words. Ask a friend to solve it.

Name _____

Skill: Choosing procedures,
Addition and subtraction with regrouping

Summer Camp Fun

Underline the addition key words (**in all**, **altogether**, or **total**) or the subtraction key words (**how many left**, **how many more**, or **how many fewer**). Solve each problem. Circle the letter of the correct number sentence. Fill in that answer.

1. 38 boys and 43 girls went to summer camp. How many children went to summer camp altogether?		a. 38 + 43 = _____ boys b. 38 + 43 = _____ children c. 43 − 38 = _____ more girls
2. The Red Group went hiking and saw 27 animals. The Green Group saw 53 animals on its hike. How many more animals did the Green Group see?		a. 53 − 27 = _____ more animals b. 27 + 53 = _____ animals c. 53 − 27 = _____ fewer animals
3. On Wednesday 75 campers went swimming. Only 58 campers went swimming on Thursday. How many fewer campers went swimming on Thursday?		a. 75 + 58 = _____ campers b. 75 − 58 = _____ fewer days c. 75 − 58 = _____ fewer campers
4. Two groups played baseball. The final score was Orange Group 19 and Blue Group 27. How many more points did the Blue Group score?		a. 27 − 19 = _____ fewer points b. 19 + 27 = _____ points c. 27 − 19 = _____ more points
5. At the Craft Center it costs 45¢ to make a bracelet and 49¢ to make an airplane. What is the total cost to make a bracelet and an airplane?		a. 45¢ + 49¢ = _____ ¢ b. 49¢ − 45¢ = _____ ¢ c. 45 + 49 = _____ crafts

Brainwork! Would you add or subtract to find out how many more campers brought sleeping bags than sheets and blankets? Write a word problem about this. Ask a friend to solve it.

© Frank Schaffer Publications, Inc.

FS-32022 Math Activities

Name_____ Skill: Estimation, Addition and subtraction

Baseball Days

When solving a problem, first estimate what the answer will be.
1. Underline the key words.
2. Circle the most reasonable answer.
3. Write and solve the number sentence for the **exact** answer.
4. Write the answer.
5. Check that the answer **makes sense**.

---

1. There are 21 boys and 30 girls in the league this summer. How many baseball players in all are in the league?

   About 10   (About 50)

   $21 + 30 = 51$

   __51__ players

---

2. Mario's team practiced throwing for 8 minutes and batting for 19 minutes. How many more minutes did they spend practicing batting?

   About 10   About 40

   _____

   _____ minutes

---

3. On Wednesday Mario's team scored 6 runs. Alicia's team scored 19 runs. How many fewer runs did Mario's team score than Alicia's?

   About 10   About 50

   _____

   _____ runs

---

4. Alicia's team practiced for 38 minutes on Friday and 60 minutes on Saturday. How many minutes did they practice altogether?

   About 50   About 100

   _____

   _____ minutes

---

5. Alicia hit the ball 57 feet on her first try and 78 feet on her next try. How many fewer feet did she hit the ball on her first try than on her second try?

   About 20   About 80

   _____

   _____ feet

---

6. Mario's team spent 23 dollars on shirts and 19 dollars on baseball hats. What was the total amount of money spent on shirts and hats?

   About 40   About 70

   _____

   _____ dollars

Brainwork! There are four bases on a baseball field. If Marco made 3 home runs, how many times did he touch a base? Estimate the answer, then solve the problem.

Name _____

Skill: Locating information, Addition and subtraction

The Snack Shop

The Snack Shop needs your help. The cash registers are not working. Fill in the prices and total up the orders. Be sure to give back the correct change. Order **A** has been done for you.

Menu
Superburger	$.89	Green Salad	$.39
Cheeseburger	$.58	Milk	$.25
Pizza Slice	$.49	Apple Juice	$.55
Fruit Cup	$.28	Yogurt	$.35

A. *Snack Shop*
1 Yogurt $.35
1 Milk $.25
1 Fruit Cup $.28
Total Cost $.88

Money received $.95
Total Cost   -$.88
Change $.07

B. *Snack Shop*
3 Yogurts   _____
            _____
            _____
Total Cost _____

Money received $1.25
Total Cost   _____
Change   _____

C. *Snack Shop*
2 Superburgers _____
               _____
1 Fruit Cup   _____
1 Green Salad _____
2 Apple Juice _____
               _____
Total Cost   _____

Money received $3.75
Total Cost   _____
Change   _____

D. *Snack Shop*
1 Superburger _____
1 Milk   _____
1 Fruit Cup   _____
Total Cost   _____

Money received $1.50
Total Cost   _____
Change   _____

E. *Snack Shop*
1 Pizza Slice _____
1 Milk   _____
1 Fruit Cup   _____
Total Cost   _____

Money received $2.00
Total Cost   _____
Change   _____

Brainwork! Order lunch for yourself. Total the amount.

© Frank Schaffer Publications, Inc.

Name _____   Skill: Missing information,
                                                Addition and subtraction

Something Is Missing

Some problems do not give all the facts you need to find the answer. Another fact is needed here:

    Chris ate 3 slices of pizza.
    Roy ate pizza, too.
    How many slices did they eat altogether?

To solve the problem, you need to know how many slices of pizza Roy ate.

These pictures supply the information needed to solve the problems below. Write the letter of the picture that helps you solve the problem. Then work each problem and write its answer.

A (Money Bank Hal 80¢)	B (3 packages for 30¢)	C (Max 70¢)	D (Ashley 75¢)	E	F (42¢)

1. Ashley bought one kite for 54¢. She spent some of her money. How much money did she have left?
   Which picture has the missing information? ____
   ____ ¢

2. Lynn used all of her money to buy 4 packages of stickers. How much money did she spend?
   Which picture has the missing information? ____
   ____ ¢

3. Rita has saved 50¢. Hal has saved money, too. How much more money has Hal saved than Rita?
   Which picture has the missing information? ____
   ____ ¢ more

4. Carla saves pennies. She has 5 pennies in each jar. How many pennies does she have in all?
   Which picture has the missing information? ____
   ____ pennies

5. On Friday Paco earned 75¢. He bought three seed packages. How much money does he have left?
   Which picture has the missing information? ____
   ____ ¢

Brainwork! Max wants to buy baseball cards with his money. Each package costs 33¢. How many packages can he buy? Find the missing information above. Then solve the problem.

© Frank Schaffer Publications, Inc.   9   FS-32022 Math Activities

Name _____ Skill: Extra information,
Addition and subtraction facts 11-18

Shopping for School Supplies

Sometimes problems have too much information. Cross out any extra information. Add or subtract to solve each problem. Then write the answer.

1. Dad took Jessica and Troy shopping for 3 hours. Jessica brought 7 quarters she had saved. Troy brought 8 quarters. How many quarters did Jessica and Troy bring in all?	_____ quarters
2. Jessica bought a sweatshirt on sale for 3 dollars. The matching slacks she bought cost 9 dollars. Dad said not to spend more than 15 dollars. How much did she spend altogether on her new outfit?	_____ dollars
3. It took Troy 14 minutes to buy a T-shirt and 6 minutes to buy socks. He looked at books for 5 minutes. How many more minutes did it take Troy to buy his new T-shirt than to buy his socks?	_____ minutes
4. Jessica bought 5 new notebooks, 16 pencils, and 7 erasers. How many more pencils than erasers did Jessica buy?	_____ more pencils
5. Jessica bought 2 boxes of markers. Each box had 9 markers in it. Troy only wanted a box of 24 crayons. How many markers did she buy in all?	_____ markers
6. Troy bought a package of pencils for $2.39. There were 13 red, 7 blue, and 8 green pencils in the package. How many more red than blue pencils were in the package?	_____ more red pencils

Brainwork! Write a subtraction word problem. Use extra information. Give your problem to a friend to solve.

© Frank Schaffer Publications, Inc.   10   FS-32022 Math Activities

Name _____   Skill: Using pictures, Multiplication

Pat's Fishbowl

Sometimes a picture can help you solve a problem. Cut and paste the correct picture beside the matching problem. Write the multiplication number sentence on the line.

1. Pat bought 3 different kinds of fish for his fishbowl. He came home with 3 of each kind. How many fish does Pat have in all?		A
2. Pat bought 2 different kinds of snails. He bought three of each kind. How many snails does Pat have altogether?		B
3. Pat used all his nickels to buy the fish. He laid out 5 rows of nickels with 4 in each row. How many nickels did he have?		C
4. Pat filled a 4-cup measuring cup 3 times to fill his fishbowl with water. How many cups of water did Pat use in his fishbowl?		D
5. Pat has 9 fish in his fishbowl. Selena has twice as many fish in her aquarium. How many fish does Selena have?		E

© Frank Schaffer Publications, Inc.   FS-32022 Math Activities

Name _____   Skill: Multiplication key words,
                                                         Multiplication facts

Insect Watching

These key words often mean that the answer can be found by multiplying. Use them to help you solve multiplication problems. Some multiplication key words are the same as addition key words.

total   times
in all   each
altogether

Underline the key words. Write the number sentence and the answer.

1. Trevor saw 3 ladybugs. Each ladybug had 5 spots. How many ladybug spots did Trevor see in all? _____ Trevor saw _____ ladybug spots.	2. Betty caught 6 fireflies. Danielle caught 4 times as many fireflies. How many fireflies did Danielle catch? _____ Danielle caught _____ fireflies.
3. Chandra saw 2 butterflies on each of the 9 daisies in her garden. How many butterflies did Chandra see? _____ Chandra saw _____ butterflies.	4. Manny watched 6 ants working. Each of the ants had 6 legs. How many ant legs did Manny count altogether? _____ Many counted _____ ant legs.
5. Rose counted 7 dragonflies at the lake. She counted 5 times as many mosquitoes. How many mosquitoes did Rose count? _____ Rose counted _____ mosquitoes.	6. Lee found 7 twigs. There were 6 moth eggs on every twig. What was the total number of moth eggs Lee found? _____ Lee found _____ moth eggs.

Brainwork! Write a multiplication word problem about grasshoppers. Use multiplication key words. Have a friend solve it.

© Frank Schaffer Publications, Inc.   FS-32022 Math Activities

Name_____   Skill: Two-step addition/subtraction,
Facts 11-18

Fun on the Playground

Sometimes it takes more than one step to solve a problem. Read the problems below. Use two steps to solve each problem.

1. In the first inning of the kickball game Emily's team scored 9 runs. They scored 5 more runs the next inning and 3 in the last. How many runs did Emily's team score in all?

   _____ _____

   _____ runs

2. There were 17 children running a race. 8 went to get a drink. Then 6 went to join the kickball game. How many were left in the running race?

   _____ _____

   _____ children

3. There were 6 children jumping rope. 9 more joined them. The line was too long so 7 went to go swing. How many children were left jumping rope?

   _____ _____

   _____ children

4. There were 14 children swinging. 7 jumped off to go to the slide. Soon 6 more children came to swing. How many children are now swinging?

   _____ _____

   _____ children

5. Linda climbed 13 steps up the ladder. She climbed down 4 steps to talk to Tara. Then she climbed up 8 steps to get to the top. How many steps high is the ladder?

   _____ _____

   _____ steps high

6. There were 16 balls on the playground. 5 of the balls were basketballs and 4 were softballs. All of the other balls were kickballs. How many kickballs were there?

   _____ _____

   _____ kickballs

Brainwork! There were 17 children sliding. Six were on the ladder and 2 were going down the slide. The rest were waiting to climb the ladder. How many children were waiting to climb?

© Frank Schaffer Publications, Inc.   13   FS-32022 Math Activities

Name _____ Skill: Two-step problems,
Addition, Subtraction, Multiplication

At the Movies

Mom and Dad took Carlos and Teresa to the movies. Use two steps to solve each problem. Write the answer.

1. Teresa bought a drink for $.45 and peanuts for $.35. Later she bought popcorn for $.55. How much money did Teresa spend altogether?

   _____ _____

   Teresa spent _____.

2. At the refreshment stand, Carlos saw 4 shelves with 5 boxes of popcorn on each shelf. A man bought 3 of the boxes. How many boxes were left?

   _____ _____

   There were _____ boxes left.

3. There were 34 people in Teresa's row. 9 of them were mothers and 18 were children. The rest were fathers. How many fathers were in the row?

   _____ _____

   There were _____ fathers.

4. Carlos bought a large popcorn for 70¢. He paid with a 50¢ piece and a quarter. How much change did Carlos receive?

   _____ _____

   Carlos received _____ change.

5. The theater can seat 292 people. One day 197 children and 88 adults came to see the show. How many theater seats were empty?

   _____ _____

   _____ seats were empty.

6. During intermission Dad bought two boxes of popcorn for $.55 each. How much change did he receive if he gave the clerk $2.00?

   _____ _____

   Dad received _____ change.

Brainwork! Write a two-step problem for a friend to solve.

Name _____ Skill: Extra information, Addition, Subtraction, Multiplication

Desk-Cleaning Day

It's Desk-Cleaning Day! Read each problem to find out all the missing things that have turned up. Cross out any extra information. Then write the number sentence and the answer.

1. John found his 2 library books. Tina found 3 times as many library books as John. Tina also found $.07. How many library books did Tina find?

   _____

   Tina found _____ library books.

2. Dirk found 12 erasers, 19 pencils, and 17 Halloween stickers. How many fewer erasers than pencils did Dirk find?

   _____

   Dirk found _____ fewer erasers.

3. Chad found $.45 under his math book and $.38 in his crayon box. He found 11 markers and 2 pens. How much money did Chad find altogether?

   _____

   Chad found _____ .

4. Julie cleaned her desk in 27 minutes. She found 13 pennies. Christy cleaned her desk in 9 minutes. How much less time did it take Christy to clean her desk than Julie?

   _____

   Christy took _____ minutes less.

5. Mona found her sticker book! She also found 18 dinosaur stickers, 29 dog stickers, and 8 flower stickers. How many animal stickers did Mona find?

   _____

   Mona found _____ animal stickers.

6. Tasha found 5 centimeter rulers, 2 quarters, and some dimes. There were 4 times as many dimes as quarters. How many dimes did Tasha find in her desk?

   _____

   Tasha found _____ dimes.

Brainwork! Josie and Vicki each have less than 24 but more than 18 markers. Together they have 38 markers and 48 crayons. How many markers do each of the girls have?

© Frank Schaffer Publications, Inc.   15   FS-32022 Math Activities

Name _____ Skill: Division facts

In the Gym

Divide to solve these problems. Write the number sentence and the answer.

1. 18 children lined up in 3 rows to do situps. How many children were in each row?

   _____

   _____ children were in each row.

2. Ms. Simms divided 36 children into 4 teams. How many children were on each team?

   _____

   _____ children were on each team.

3. There are 6 tumbling mats. Ms. Simms divided 24 children into equal groups for each mat. How many children were in each group?

   _____

   _____ children were in each group.

4. 16 girls want to shoot baskets. Mr. Young will have the same number of girls at all 8 baskets. How many girls will be at each basket?

   _____

   _____ girls will be at each basket.

5. 3 equal groups of children want to run relay races. There are 21 children. How many children are in each group?

   _____

   _____ children are in each group.

6. 32 boys lined up in 4 equal rows to do jumping jacks. How many boys were in each row?

   _____

   _____ boys were in each row.

7. At the end of class, 27 kickballs must be put equally into 3 bags. How many kickballs fit in each bag?

   _____

   _____ kickballs fit in each bag.

Brainwork! Tom's dog eats 3 bones each day. How long will a box of 18 bones last? Draw pictures to prove your answer.

© Frank Schaffer Publications, Inc. 16 FS-32022 Math Activities

Name _____   Skill: Understanding a bar graph

Speeds of Animals

Here is a bar graph that shows the speeds of some animals. Use the graph to answer the questions below.

Miles Per Hour (m.p.h.)

Cheetah	70
Lion	50
Zebra	40
Rabbit	35
Grizzly Bear	30
House Cat	30
Elephant	25
Wild Turkey	15

1. Which animal can run 70 m.p.h.? _____

2. How fast can a lion run? _____

3. Which two animals can run the same speed? _____
_____

4. Which is faster, a grizzly bear or a zebra? _____

5. Which of these animals is the slowest, a lion, zebra, cheetah, or rabbit? _____

6. How much faster is a zebra than an elephant? _____

7. A cheetah can run twice as fast as which animal? _____

8. How much slower is a house cat than a lion? _____

Brainwork! Make a list of ten different animals. Give five friends one minute to memorize all the words. After one minute, ask them to say the words. Draw a bar graph to show how many words they remembered.

© Frank Schaffer Publications, Inc.   FS-32022 Math Activities

Name _____   Skill: Logical thinking

Think December

Fill in the dates on this December calendar. Use these clues to help.

- The last day in November was on a Wednesday.
- Winter begins on Wednesday, December 21st.
- December has 31 days.

Now use the calendar and the clues below to find the answers.

December						
Sun.	Mon.	Tues.	Wed.	Thur.	Fri.	Sat.

1. If today is Monday, what day of the week will it be eight days from now?   _____

2. December 1 is on which day of the week?   _____

3. What is the date of the third Tuesday in the month?   _____

4. Nathan has drum lessons every Wednesday. How many lessons will he have in December?   _____

5. It snowed 9 days in December. How many days did it **not** snow?   _____

6. Michelle can go shopping on any Friday in December. But her mother can go with her only on December 5, 14, 16, or 20. What day should Michelle plan to go shopping with her mother?   _____

7. Vacation begins on December 23. Annie's birthday is four days earlier. What is the date of Annie's birthday?   _____

8. January 1 will be on what day of the week?   _____

Brainwork! Use the calendar above to find this date. It is on a weekend. It is an even number and has two digits. The two digits add up to six.

© Frank Schaffer Publications, Inc.   FS-32022 Math Activities

Name _____ Skill: Logical thinking

Oodles of Tadpoles

Use the clues to find out how many tadpoles each student has.

1. Yolanda has an even number of tadpoles. If you take out 4, there will be 6 in her jar. How many tadpoles does Yolanda have?

   _____ tadpoles

2. Greg has fewer than 29 tadpoles. He has more than 25. He has an odd number. How many tadpoles does Greg have?

   _____ tadpoles

3. Pablo has an odd number of tadpoles. If you double his number, he will have 18. How many tadpoles does Pablo have?

   _____ tadpoles

4. Megan has 20 tadpoles in her jar. Danny has twice as many tadpoles in his jar. How many tadpoles does Danny have?

   _____ tadpoles

5. Kathy has more than 7 but fewer than 15 tadpoles. If she counts them by fives, there is one left over. How many tadpoles does Kathy have?

   _____ tadpoles

6. Paul has more than 50 tadpoles. He has fewer than 75. The two digits in the number are the same. The sum of the digits is 12. How many tadpoles does Paul have?

   _____ tadpoles

7. Todd has more than 30 but less than 40 tadpoles. You say the number when you count by twos and when you count by threes. How many tadpoles does Todd have?

   _____ tadpoles

Brainwork! Tadpoles each have two eyes and one tail. How many tadpoles does Craig have in his jar? Hint: There are 4 more eyes than tails.

Name _____

Skill: Mixed strategies

Computer Fun

Andrew and Gina have been playing computer games. Solve each problem. Write the answer.

1. Andrew scored 48 points in Math Magic. Gina scored 75 points. How many fewer points did Andrew score than Gina?

   Andrew scored _____ fewer points.

2. Gina played Spelling Invaders 5 times. Her highest score was 49. Andrew's highest score was 68. How much higher was Andrew's highest score than Gina's?

   Andrew's score was _____ points higher.

3. Gina's score in Punctuation Trouble was less than 250 but more than 125. It is an odd number. The three digits are 2, 5, and 2. What was Gina's score?

   Gina's score was _____.

4. In Fraction Frogger Andrew had 60 points. Then he lost 35 points. Finally he scored 55 more points for getting Frog across the river. What was Paul's final score?

   Paul's final score was _____.

5. Gina saw a pattern in her Amazing Reader scores. She scored 68, 78, 88, 98, 108, and 118. Explain the pattern she saw.

   _____
   _____
   _____

6. Andrew played Math Magic 3 times. Gina played it 8 times as often as Andrew. How many games of Math Magic did Gina play? Circle the letter of the correct number sentence and answer.

   a. 3 + 8 = 11 games
   b. 8 x 3 = 24 games
   c. 8 x 3 = 24 points

Brainwork! Write two word problems about computer games. Ask a friend to solve them.

Name _____

Skill: Multiplying by 2

2	4	6	8	10
blue	red	orange	yellow	green

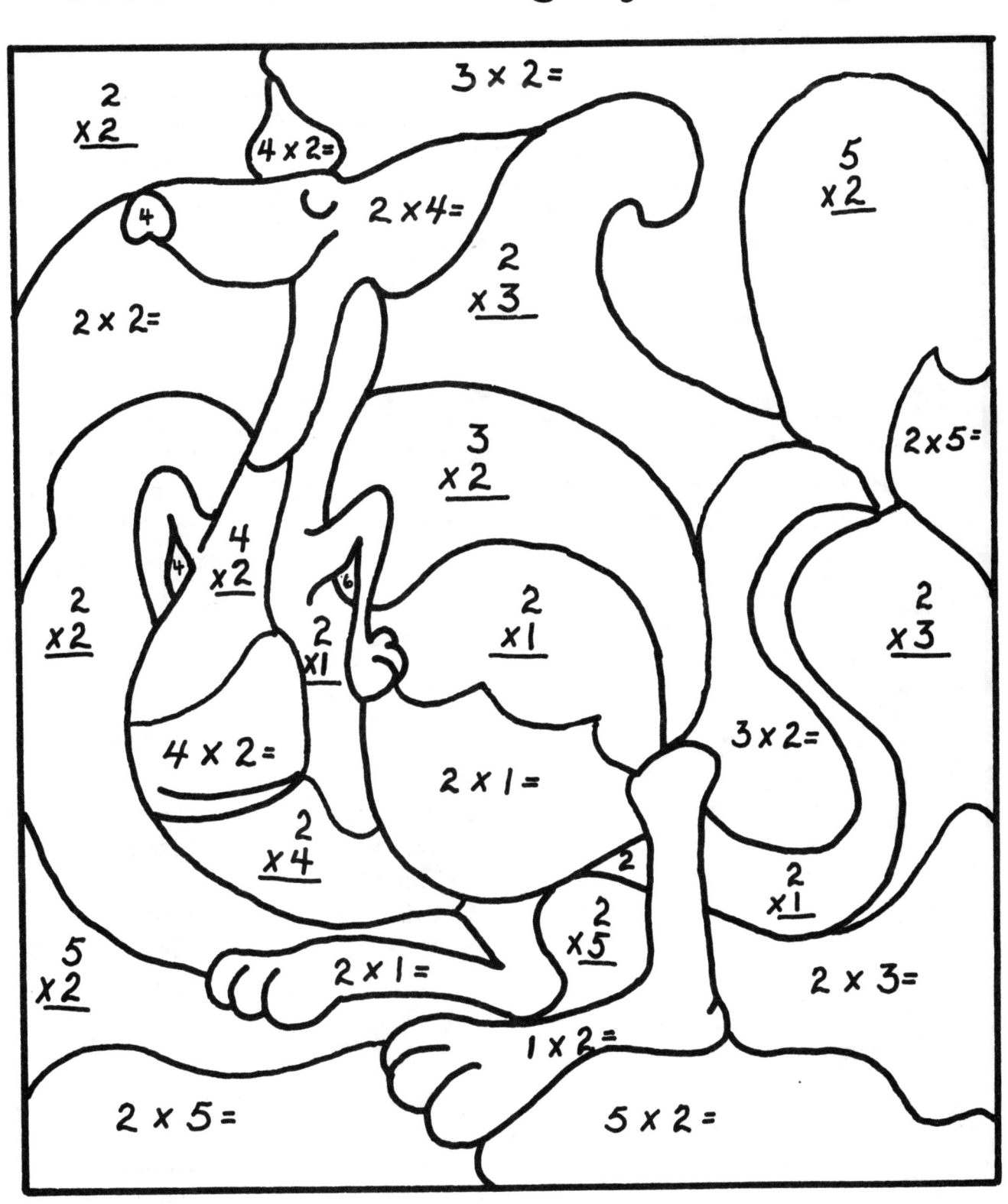

$\begin{array}{r}2\\ \times 2\\ \hline\end{array}$

$3 \times 2 =$

$4 \times 2 =$

$2 \times 4 =$

$\begin{array}{r}5\\ \times 2\\ \hline\end{array}$

$2 \times 2 =$

$\begin{array}{r}2\\ \times 3\\ \hline\end{array}$

$\begin{array}{r}3\\ \times 2\\ \hline\end{array}$

$2 \times 5 =$

$\begin{array}{r}2\\ \times 2\\ \hline\end{array}$

$\begin{array}{r}4\\ \times 2\\ \hline\end{array}$

$\begin{array}{r}2\\ \times 1\\ \hline\end{array}$

$\begin{array}{r}2\\ \times 1\\ \hline\end{array}$

$\begin{array}{r}2\\ \times 3\\ \hline\end{array}$

$4 \times 2 =$

$2 \times 1 =$

$3 \times 2 =$

$\begin{array}{r}2\\ \times 4\\ \hline\end{array}$

$\begin{array}{r}2\\ \times 5\\ \hline\end{array}$

$\begin{array}{r}2\\ \times 1\\ \hline\end{array}$

$\begin{array}{r}5\\ \times 2\\ \hline\end{array}$

$2 \times 1 =$

$1 \times 2 =$

$2 \times 3 =$

$2 \times 5 =$

$5 \times 2 =$

© Frank Schaffer Publications, Inc.

21

FS-32022 Math Activities

Name _____

Skill: Multiplying by 3

3	6	9	12	15
blue	red	orange	green	yellow

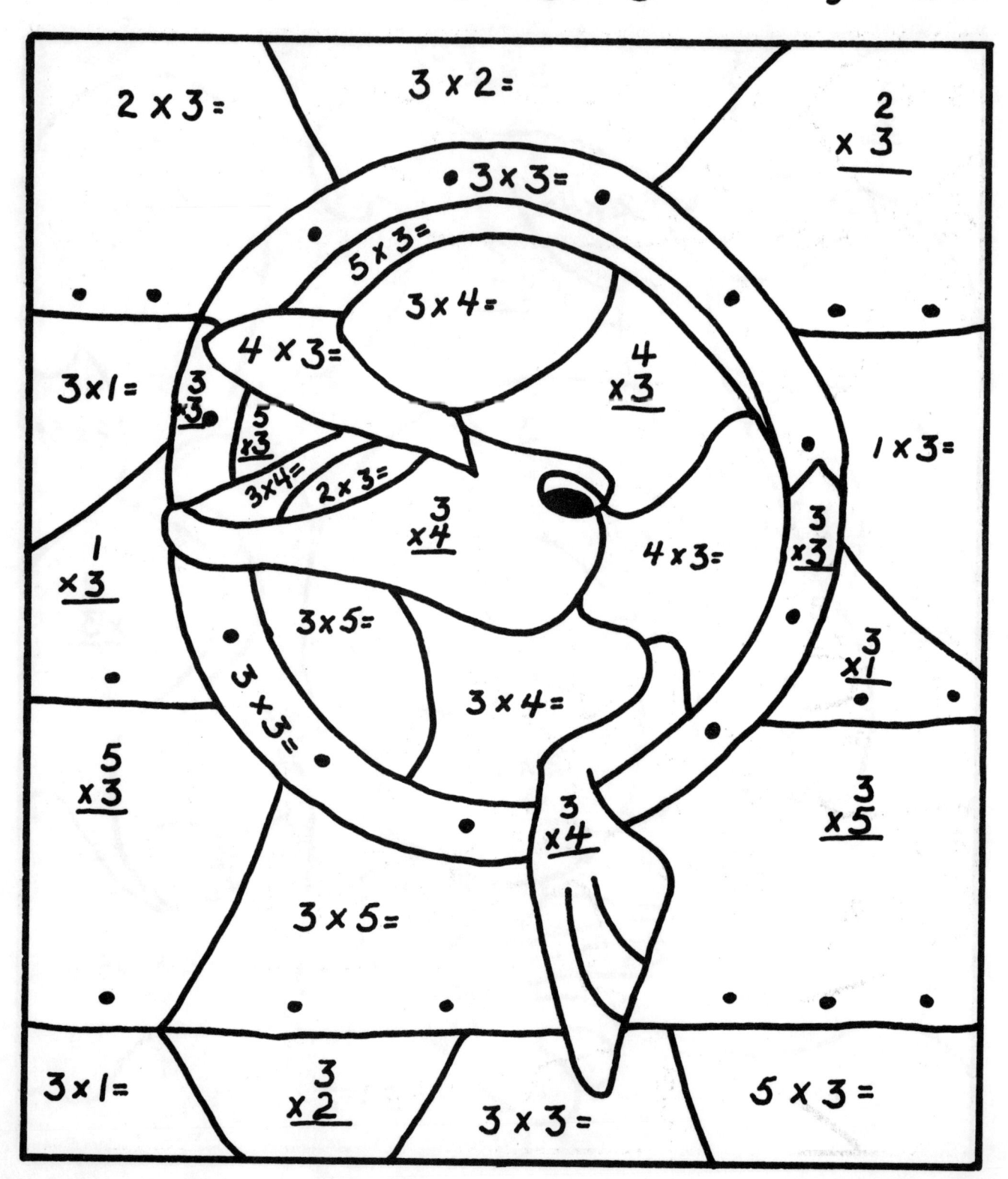

2 × 3 =

3 × 2 =

$\begin{array}{r}2\\ \times 3\\ \hline\end{array}$

3 × 3 =

5 × 3 =

3 × 4 =

4 × 3 =

$\begin{array}{r}4\\ \times 3\\ \hline\end{array}$

3 × 1 =

$\begin{array}{r}3\\ \times 3\\ \hline\end{array}$

$\begin{array}{r}5\\ \times 3\\ \hline\end{array}$

1 × 3 =

3 × 4 =

2 × 3 =

$\begin{array}{r}3\\ \times 4\\ \hline\end{array}$

4 × 3 =

$\begin{array}{r}3\\ \times 3\\ \hline\end{array}$

$\begin{array}{r}1\\ \times 3\\ \hline\end{array}$

3 × 5 =

3 × 4 =

$\begin{array}{r}3\\ \times 1\\ \hline\end{array}$

3 × 3 =

$\begin{array}{r}5\\ \times 3\\ \hline\end{array}$

$\begin{array}{r}3\\ \times 4\\ \hline\end{array}$

$\begin{array}{r}3\\ \times 5\\ \hline\end{array}$

3 × 5 =

3 × 1 =

$\begin{array}{r}3\\ \times 2\\ \hline\end{array}$

3 × 3 =

5 × 3 =

© Frank Schaffer Publications, Inc.

FS-32022 Math Activities

Name _____

Skill: Multiplying by 3

18	21	24	27	30
orange	brown	yellow	blue	green

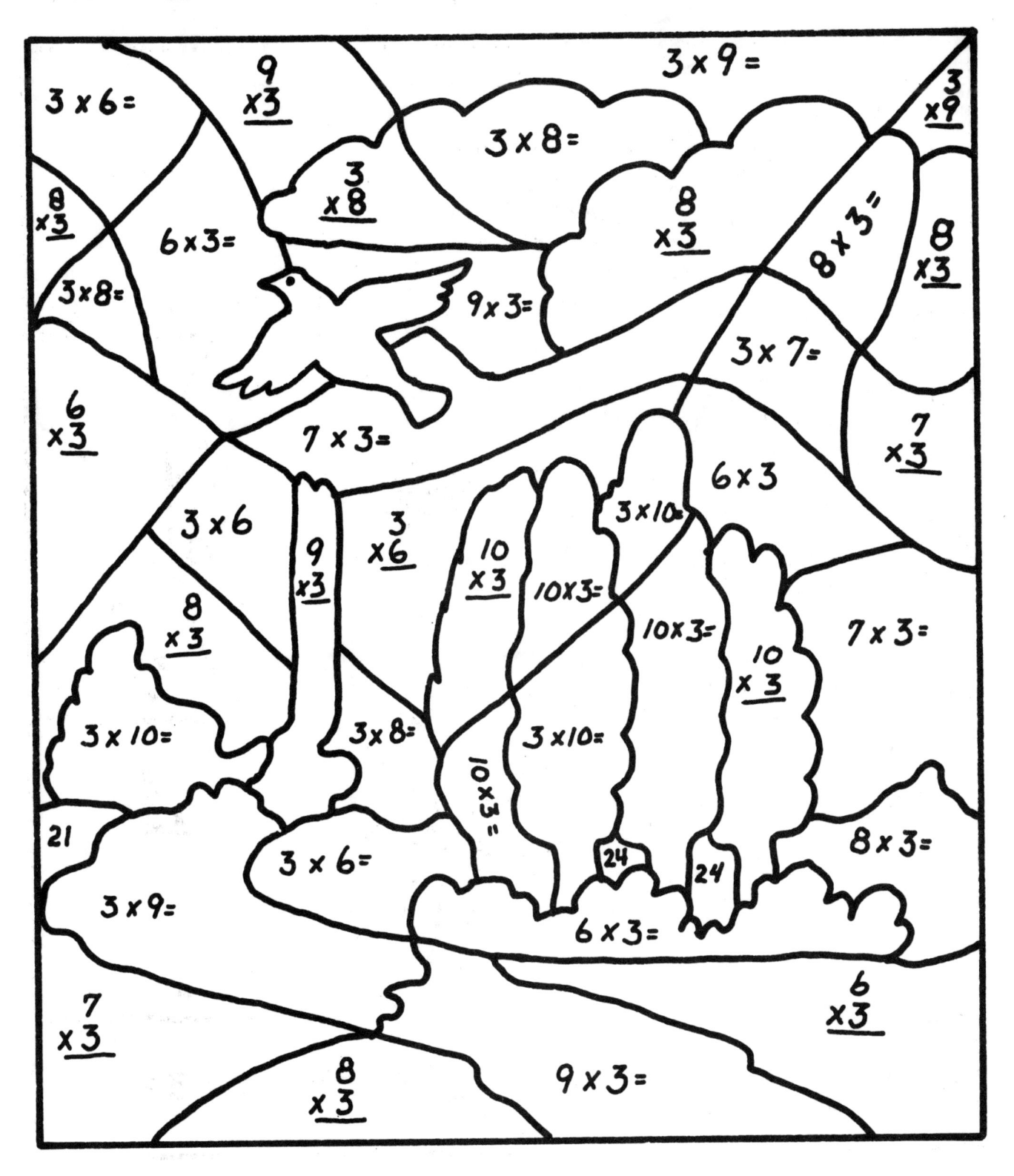

Name _____

Skill: Multiplying by 4

4	8	12	16	20
brown	red	orange	blue	yellow

Name _____

Skill: Multiplying by 4

24	28	32	36	40
yellow	blue	brown	red	orange

© Frank Schaffer Publications, Inc.

25

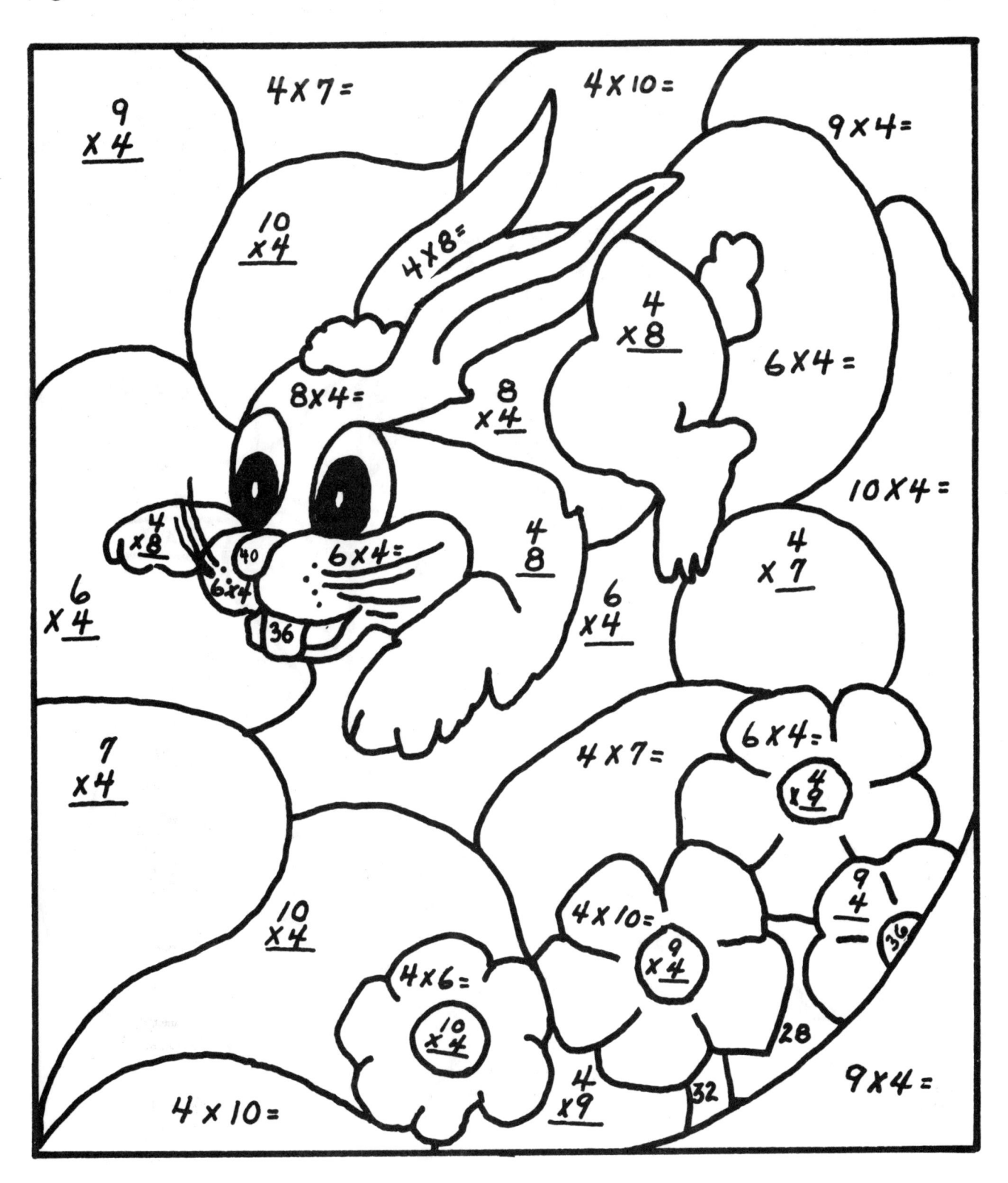

FS-32022 Math Activities

Name _____

Skill: Multiplying by 5

30	35	40	45	50
green	red	yellow	black	orange

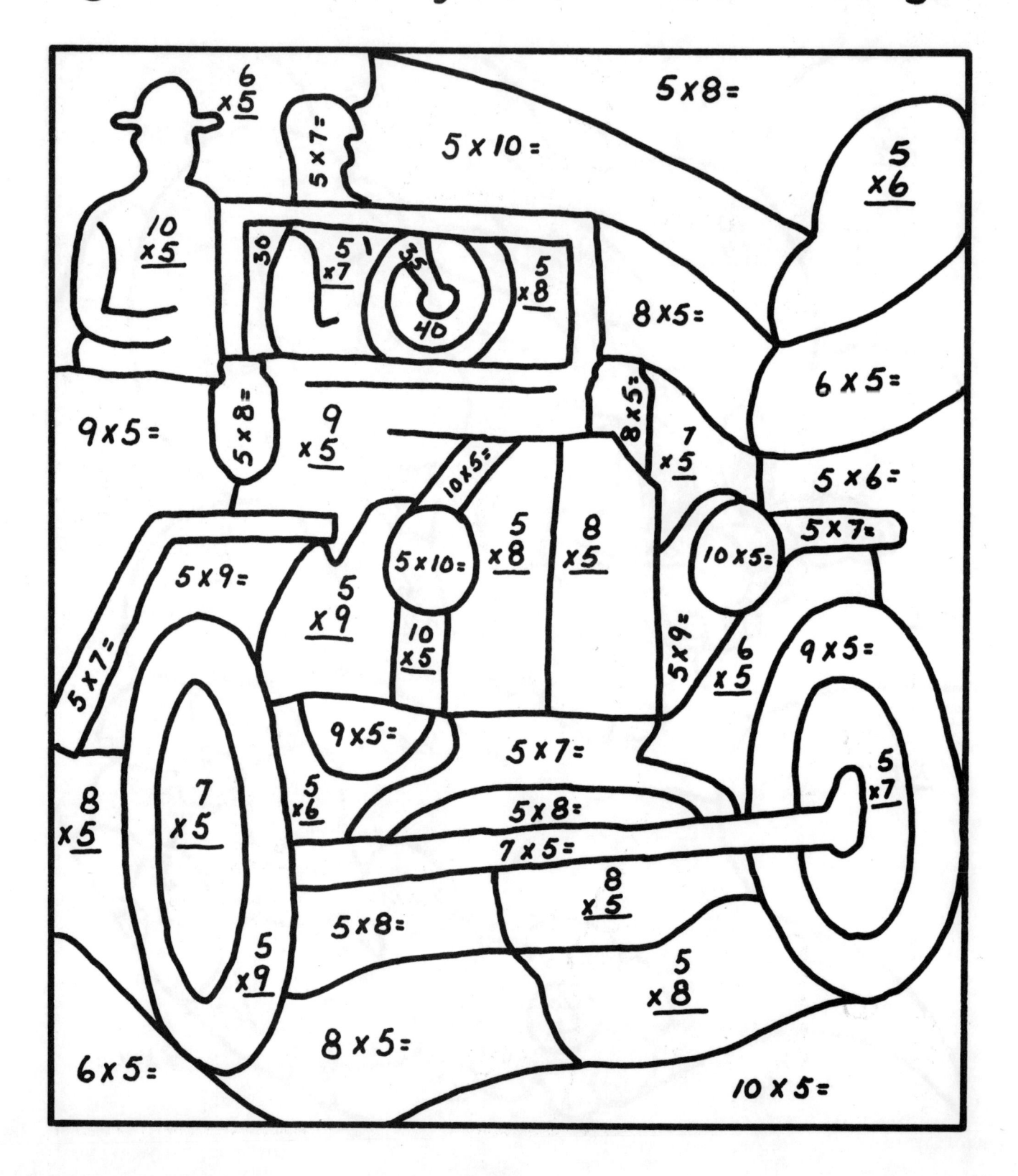

Name _____

Skill: Multiplying by 7

7	14	21	28	35
yellow	blue	red	orange	black

Name _____

Skill: Multiplying by 7

42	49	56	63	70
orange	red	yellow	black	green

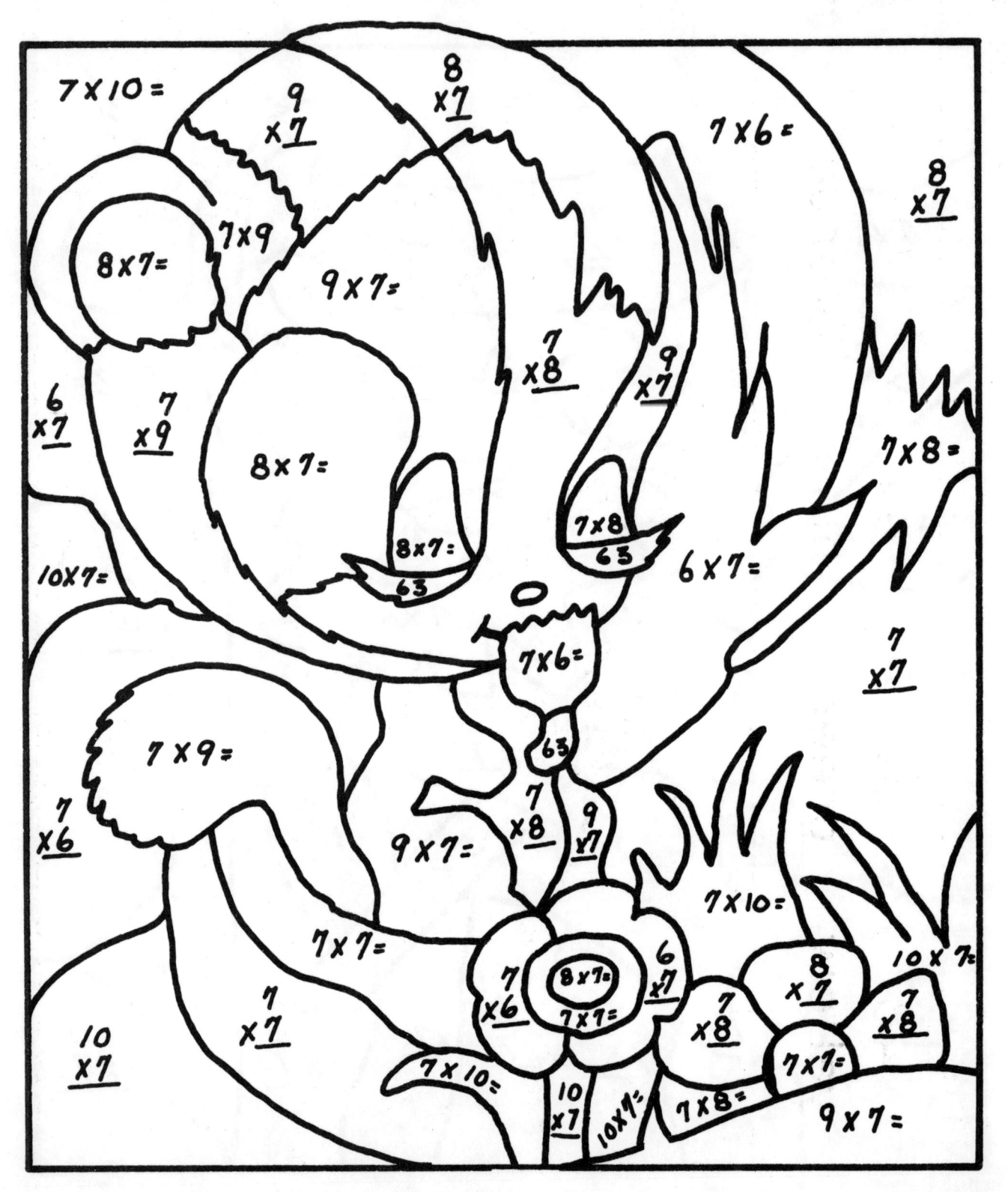

30

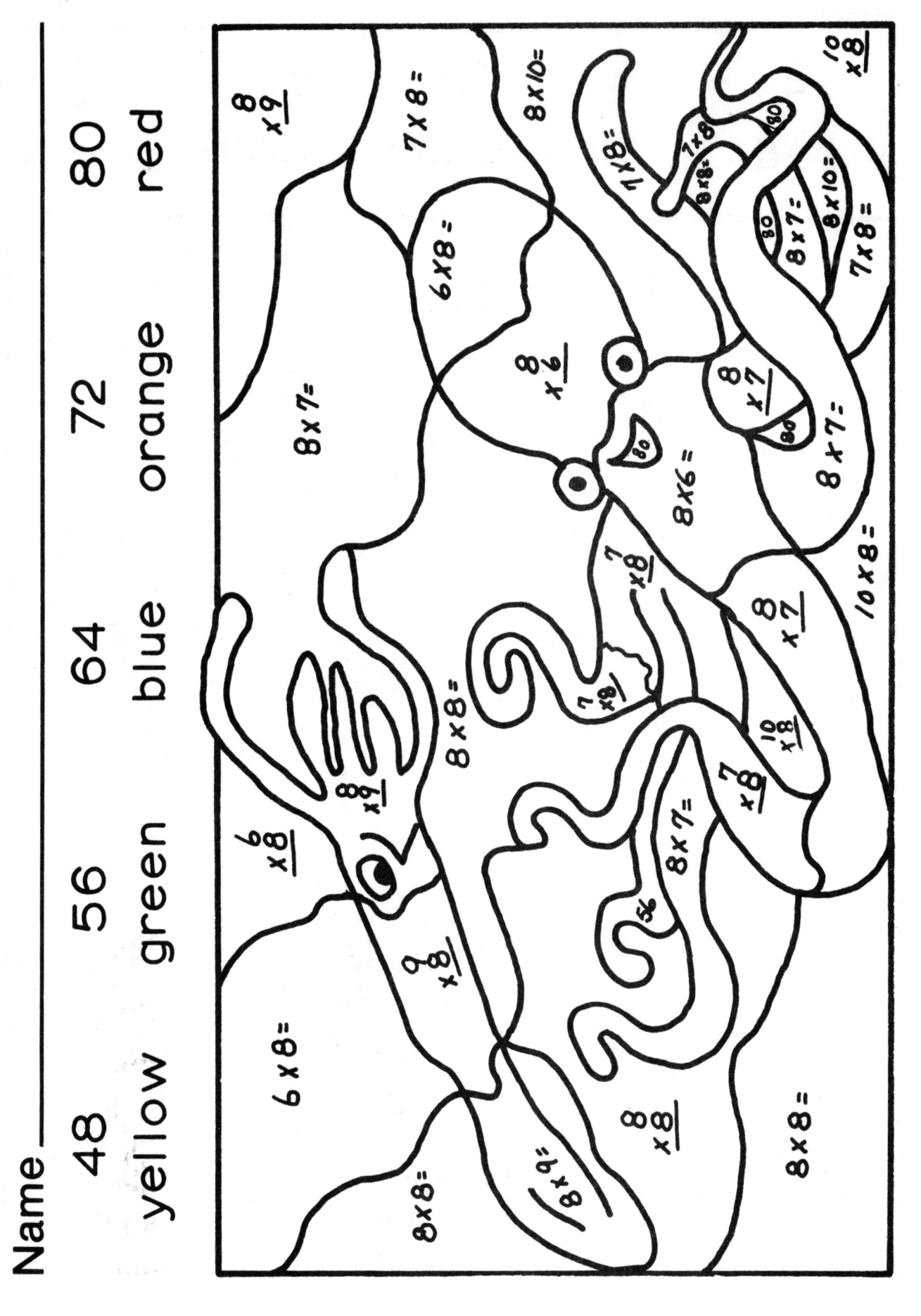

Name _____

Skill: Multiplying by 8

48	56	64	72	80
yellow	green	blue	orange	red

Name _____

Skill: Multiplying by 9

9	18	27	36	45
red	yellow	brown	green	orange

Name _____

Skill: Multiplying by 9

54	63	72	81	90
yellow	green	red	blue	orange

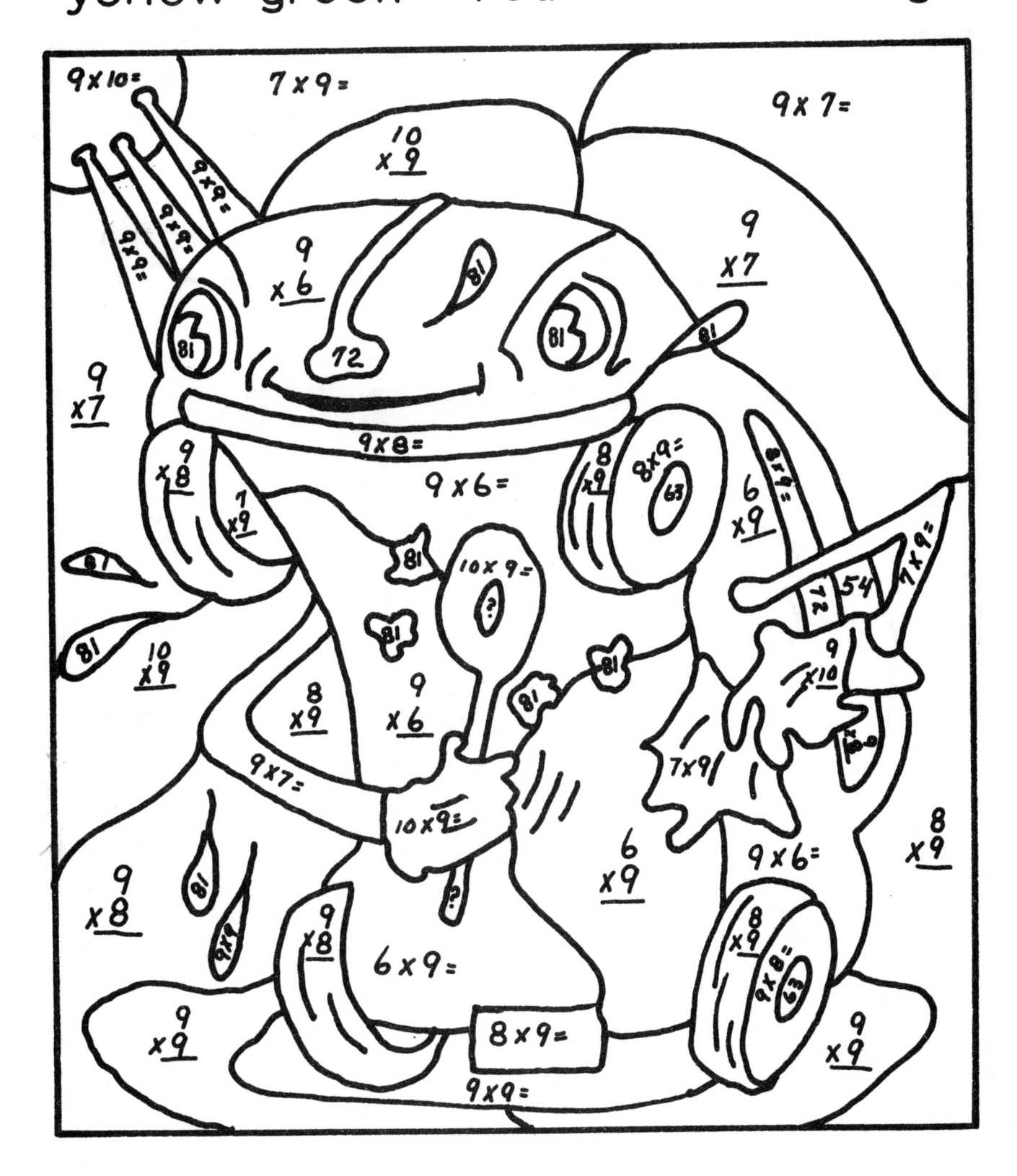

Name _____

Skill: Multiplying by 10

10	20	30	40	50
red	green	orange	brown	yellow

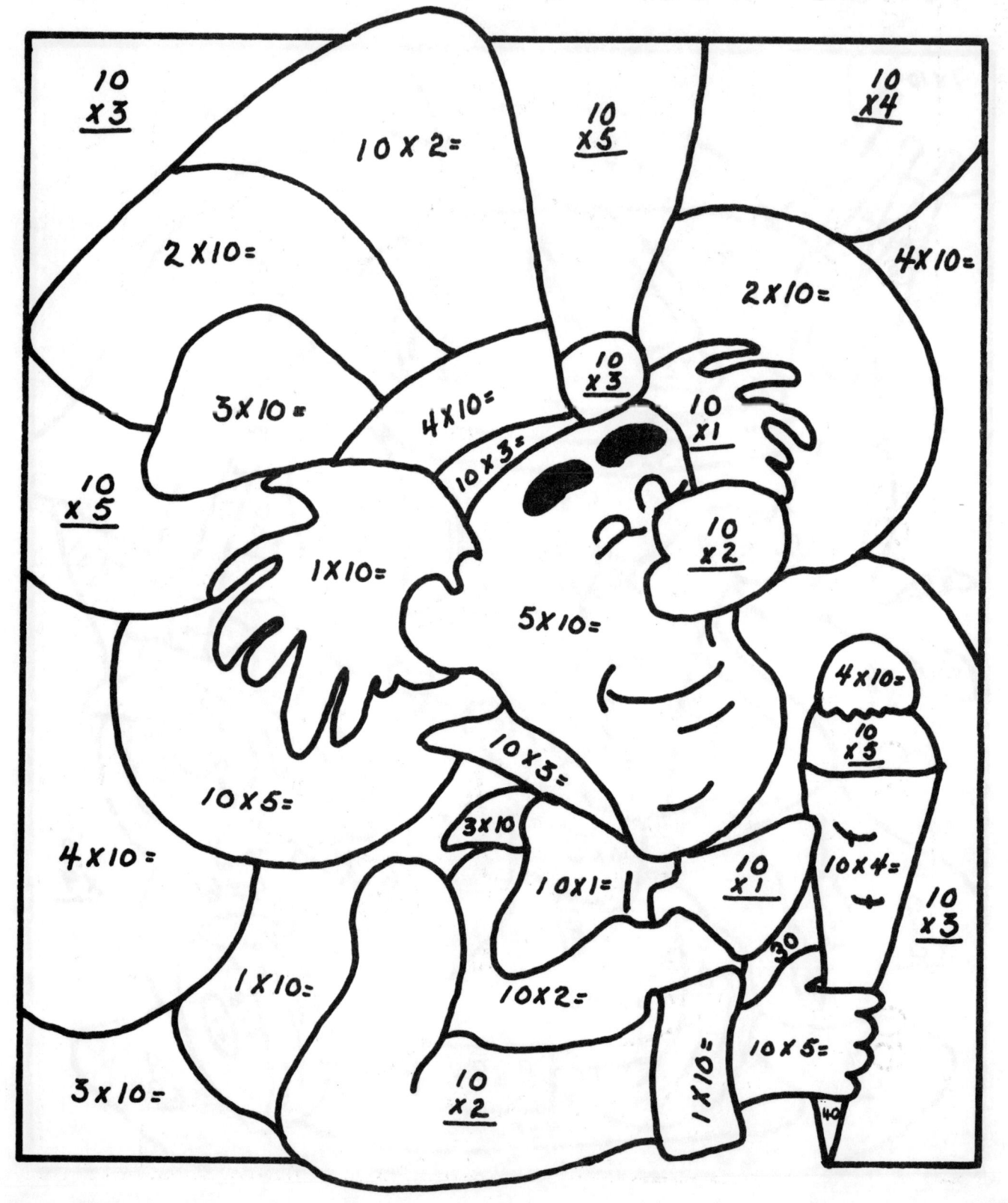

© Frank Schaffer Publications, Inc.

34

FS-32022 Math Activities

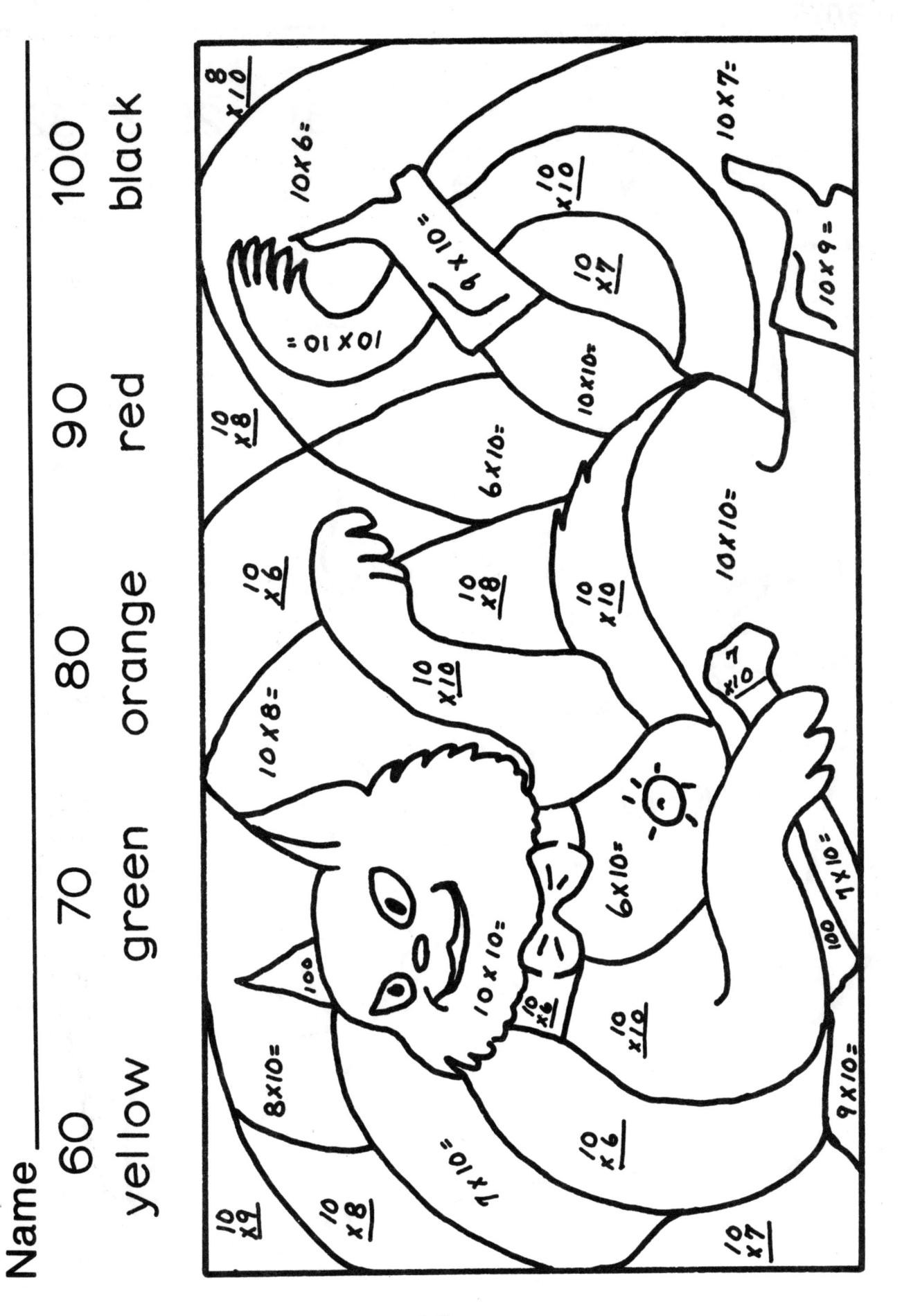

Name _____

Skill: Multiplying by 10

60	70	80	90	100
yellow	green	orange	red	black

Name _____

9	14	18	25, 27, 32
yellow	blue	brown	green

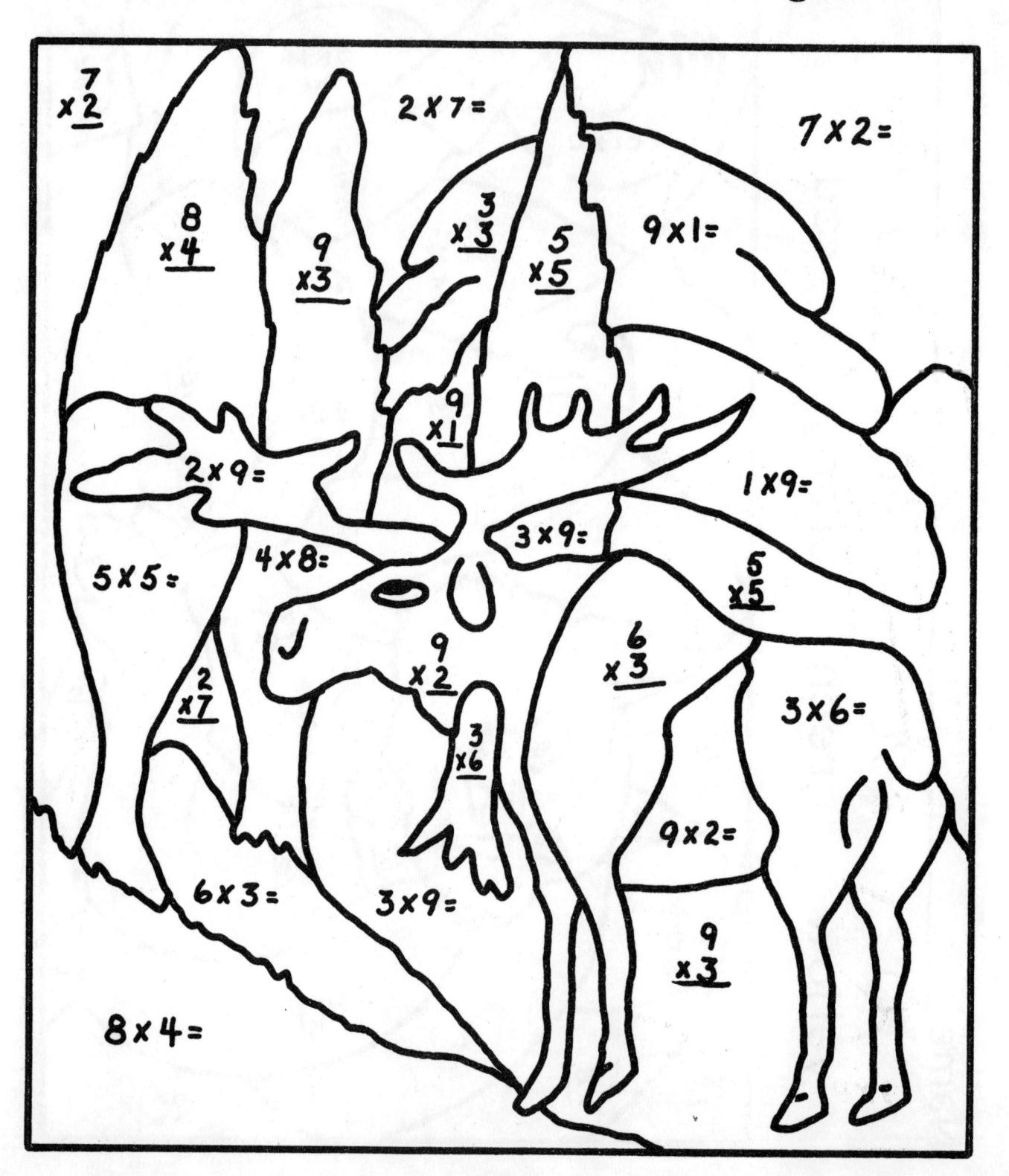

7 × 2

2 × 7 =

7 × 2 =

8 × 4

9 × 3

3 × 3

5 × 5

9 × 1 =

9 × 1

2 × 9 =

1 × 9 =

5 × 5 =

4 × 8 =

3 × 9 =

6 × 5

2 × 7

9 × 2

6 × 3

3 × 6 =

3 × 6

9 × 2 =

6 × 3 =

3 × 9 =

9 × 3

8 × 4 =

36

Name _____

20	21	36	40, 48, 54
blue	yellow	orange	green

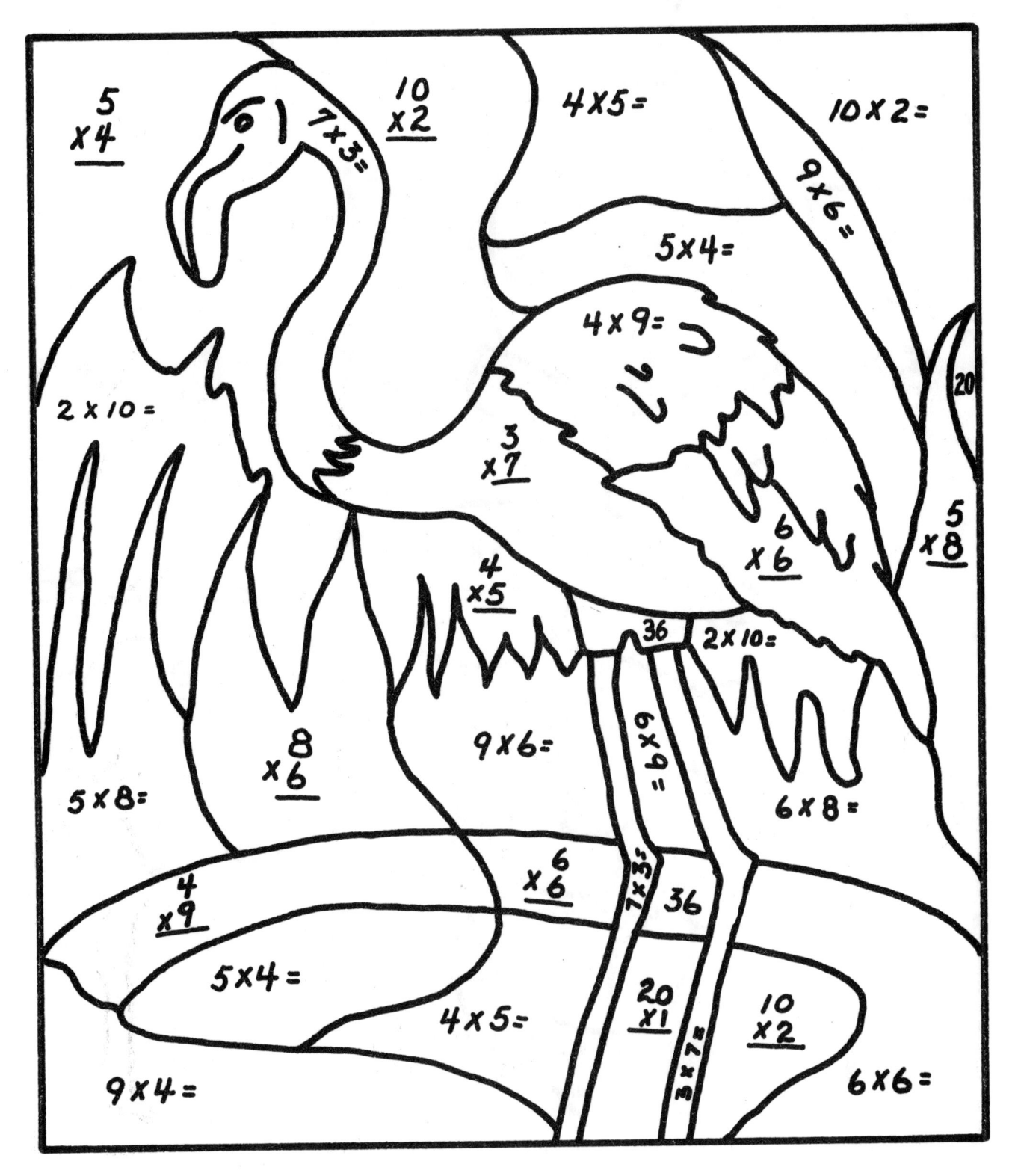

5 × 4

10 × 2

7 × 3 =

4 × 5 =

10 × 2 =

9 × 6 =

5 × 4 =

4 × 9 =

2 × 10 =

3 × 7

6 × 6

5 × 8

4 × 5

36

2 × 10 =

8 × 6

9 × 6 =

9 × 6 =

6 × 8 =

5 × 8 =

4 × 9

6 × 6

7 × 3 =

36

5 × 4 =

4 × 5 =

20 × 1

10 × 2

3 × 7 =

9 × 4 =

6 × 6 =

20

37

Name _____

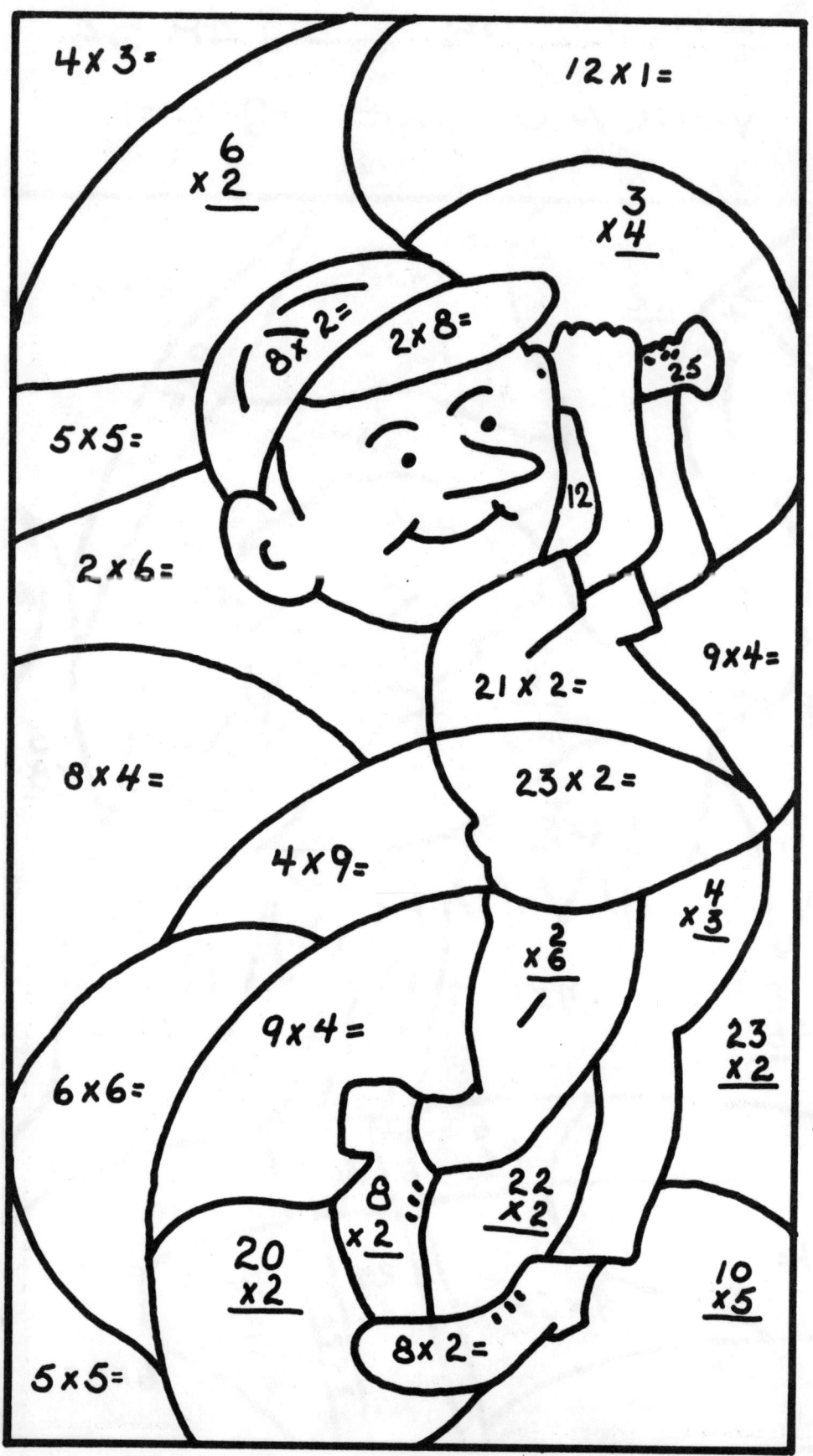

$4 \times 3 =$

$\begin{array}{r}6\\ \times\ 2\\ \hline\end{array}$

$12 \times 1 =$

$\begin{array}{r}3\\ \times\ 4\\ \hline\end{array}$

$8 \times 2 =$
$2 \times 8 =$

$5 \times 5 =$

$2 \times 6 =$

$21 \times 2 =$

$9 \times 4 =$

$8 \times 4 =$

$23 \times 2 =$

$4 \times 9 =$

$\begin{array}{r}4\\ \times\ 3\\ \hline\end{array}$

$\begin{array}{r}\times\ 2\\ \times\ 6\\ \hline\end{array}$

$9 \times 4 =$

$\begin{array}{r}23\\ \times\ 2\\ \hline\end{array}$

$6 \times 6 =$

$\begin{array}{r}8\\ \times\ 2\\ \hline\end{array}$

$\begin{array}{r}22\\ \times\ 2\\ \hline\end{array}$

$\begin{array}{r}20\\ \times\ 2\\ \hline\end{array}$

$\begin{array}{r}10\\ \times\ 5\\ \hline\end{array}$

$8 \times 2 =$

$5 \times 5 =$

12 blue

16 black

25, 32, 36 yellow

40 to 50 orange

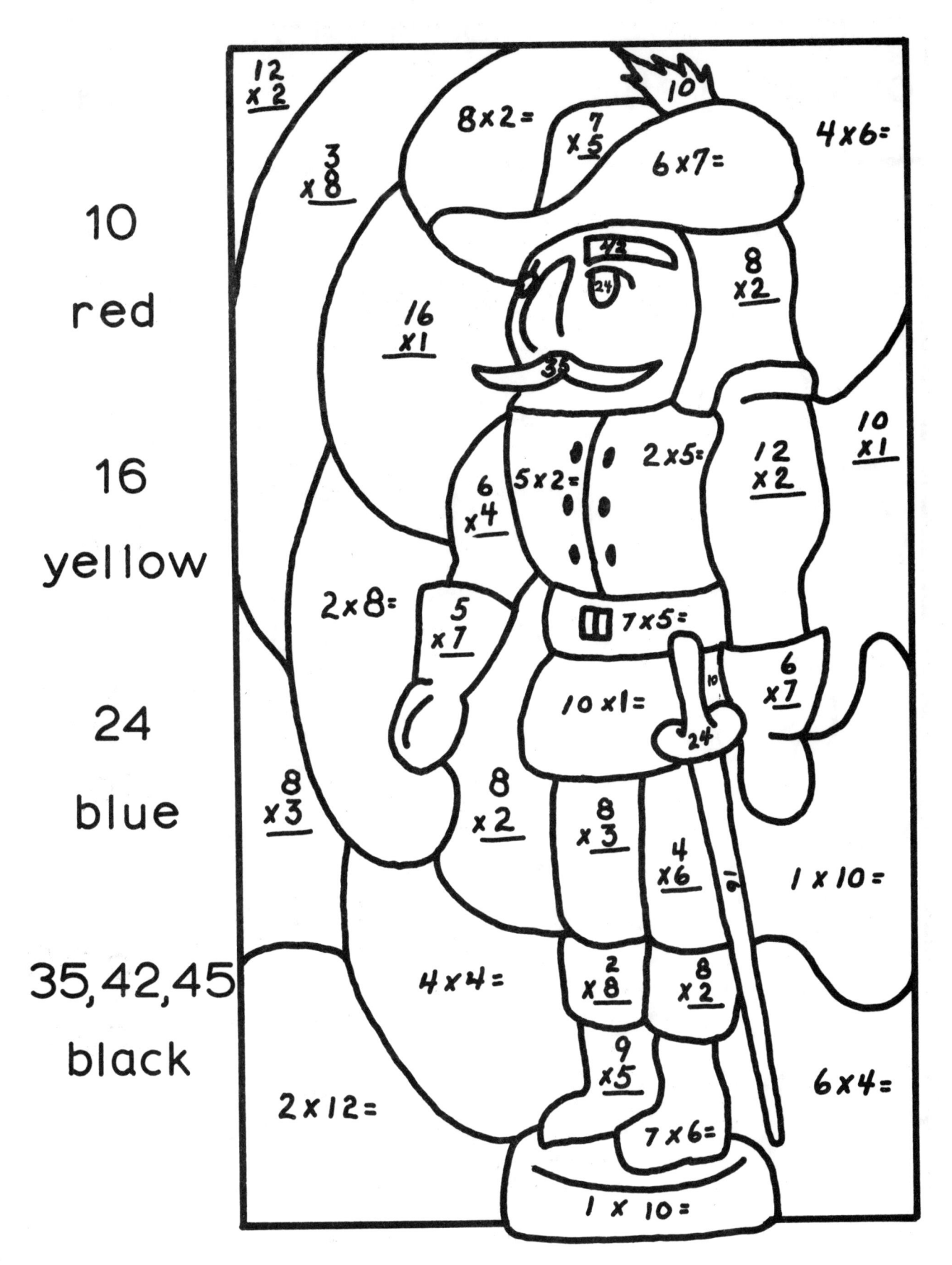

Name _____

10 red

16 yellow

24 blue

35, 42, 45 black

$\begin{array}{r}12\\\times 2\\\hline\end{array}$

$8\times 2=$

$\begin{array}{r}7\\\times 5\\\hline\end{array}$

$4\times 6=$

$\begin{array}{r}3\\\times 8\\\hline\end{array}$

$6\times 7=$

$\begin{array}{r}16\\\times 1\\\hline\end{array}$

$\begin{array}{r}8\\\times 2\\\hline\end{array}$

$5\times 2=$ $2\times 5=$

$\begin{array}{r}6\\\times 4\\\hline\end{array}$

$\begin{array}{r}12\\\times 2\\\hline\end{array}$

$\begin{array}{r}10\\\times 1\\\hline\end{array}$

$2\times 8=$

$\begin{array}{r}5\\\times 7\\\hline\end{array}$

$7\times 5=$

$10\times 1=$

$\begin{array}{r}6\\\times 7\\\hline\end{array}$

$\begin{array}{r}8\\\times 3\\\hline\end{array}$

$\begin{array}{r}8\\\times 2\\\hline\end{array}$

$\begin{array}{r}8\\\times 3\\\hline\end{array}$

$\begin{array}{r}4\\\times 6\\\hline\end{array}$

$1\times 10=$

$4\times 4=$

$\begin{array}{r}2\\\times 8\\\hline\end{array}$

$\begin{array}{r}8\\\times 2\\\hline\end{array}$

$\begin{array}{r}9\\\times 5\\\hline\end{array}$

$2\times 12=$

$7\times 6=$

$6\times 4=$

$1\times 10=$

39

Name _____

18	48	39, 42, 45	20 to 35
orange	yellow	blue	red

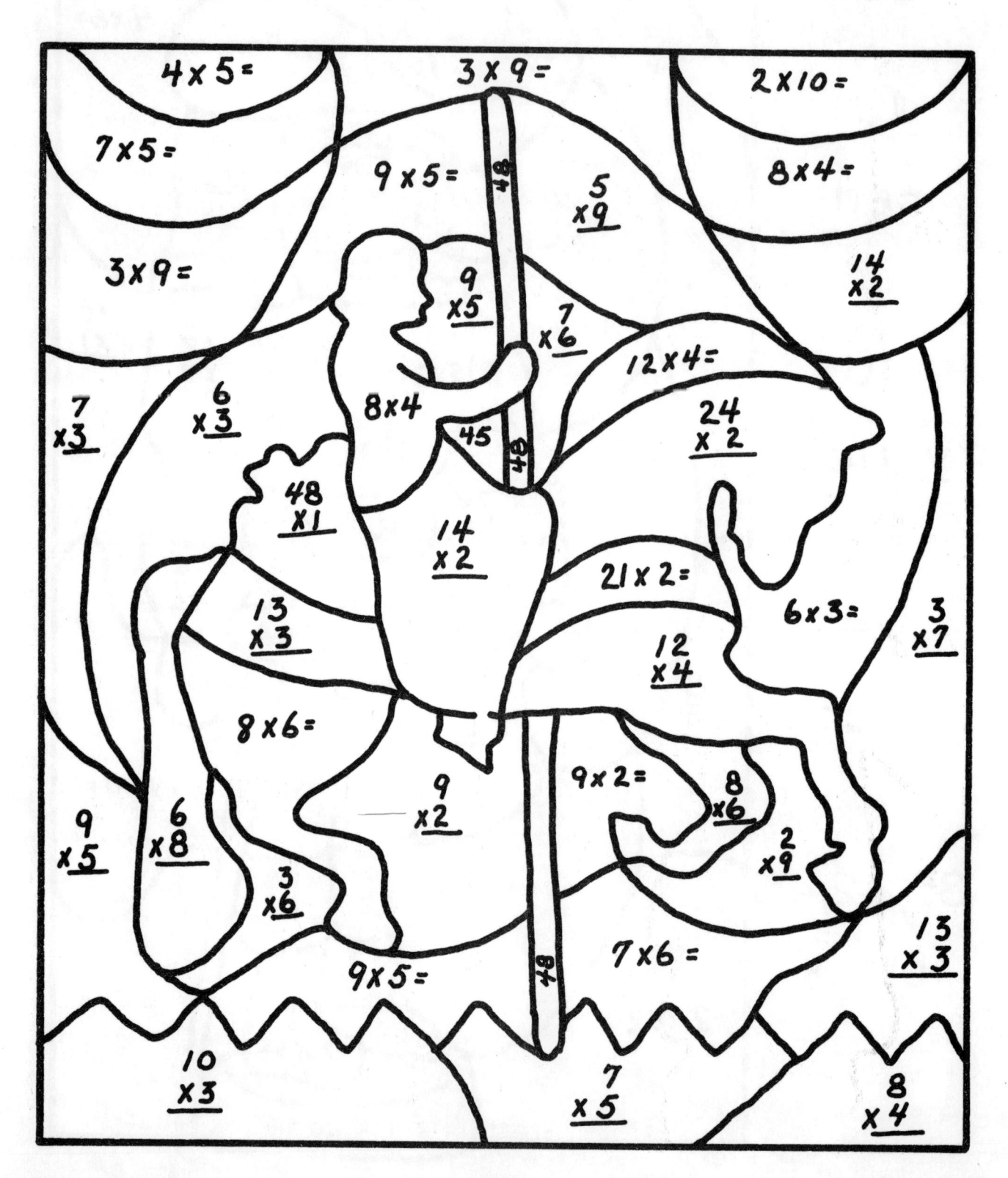

4 × 5 =   3 × 9 =   2 × 10 =
7 × 5 =   9 × 5 =   8 × 4 =
3 × 9 =   5 × 9   14 × 2
7 × 3   6 × 3   9 × 5   7 × 6   12 × 4 =
       8 × 4   45   24 × 2
       48 × 1   14 × 2   21 × 2 =   6 × 3 =   3 × 7
       13 × 3   12 × 4
8 × 6 =   9 × 2 =   8 × 6
9 × 5   6 × 8   9 × 2   2 × 9
    3 × 6   13 × 3
        9 × 5   7 × 6 =
10 × 3   7 × 5   8 × 4

40

Name _____

18	27	42, 49, 54	30 to 40
yellow	orange	brown	green

$3 \times 12 =$
10×4
10×3
11×3
9×2
6×3
4×8
10×4
12×3
9×3
9×6
$9 \times 6 =$
13×3
27
$6 \times 5 =$
6×9
7×6
7×6
7×7
$6 \times 3 =$
$6 \times 3 =$
7×7
8×4
3×6
$4 \times 10 =$
6×5
42
8×5
49
9×2
$2 \times 9 =$

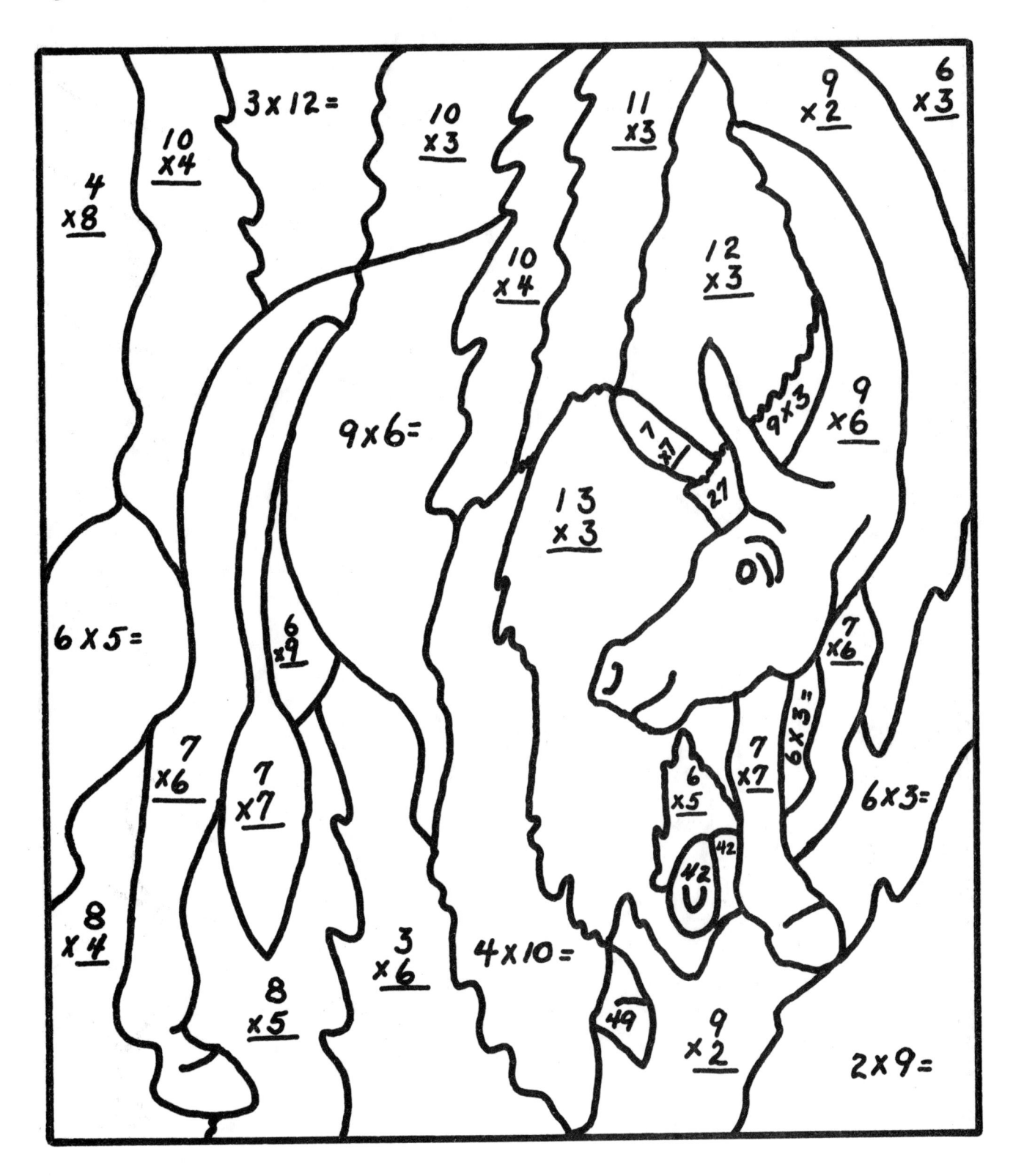

41

Name _____

30	36	49, 54, 56	60 to 75
green	yellow	black	blue

$$\begin{array}{r}23\\\times 3\\\hline\end{array}$$
$$\begin{array}{r}32\\\times 2\\\hline\end{array}$$
$$\begin{array}{r}34\\\times 2\\\hline\end{array}$$
$$\begin{array}{r}6\\\times 5\\\hline\end{array}$$
$$\begin{array}{r}31\\\times 2\\\hline\end{array}$$
$$\begin{array}{r}10\\\times 3\\\hline\end{array}$$
$7\times 8=$
$$\begin{array}{r}9\\\times 4\\\hline\end{array}$$
$9\times 7=$
$$\begin{array}{r}20\\\times 3\\\hline\end{array}$$
$3\times 10=$
$$\begin{array}{r}12\\\times 3\\\hline\end{array}$$
$6\times 9=$
$$\begin{array}{r}9\\\times 6\\\hline\end{array}$$
$7\times 7=$
$$\begin{array}{r}22\\\times 3\\\hline\end{array}$$
36
$8\times 7=$
$$\begin{array}{r}34\\\times 2\\\hline\end{array}$$
$12\times 3=$
$9\times 6=$
$6\times 6=$
$3\times 12=$
$7\times 7=$
$$\begin{array}{r}21\\\times 3\\\hline\end{array}$$
$$\begin{array}{r}4\\\times 9\\\hline\end{array}$$
$9\times 4=$

© Frank Schaffer Publications, Inc.   42   FS-32022 Math Activities

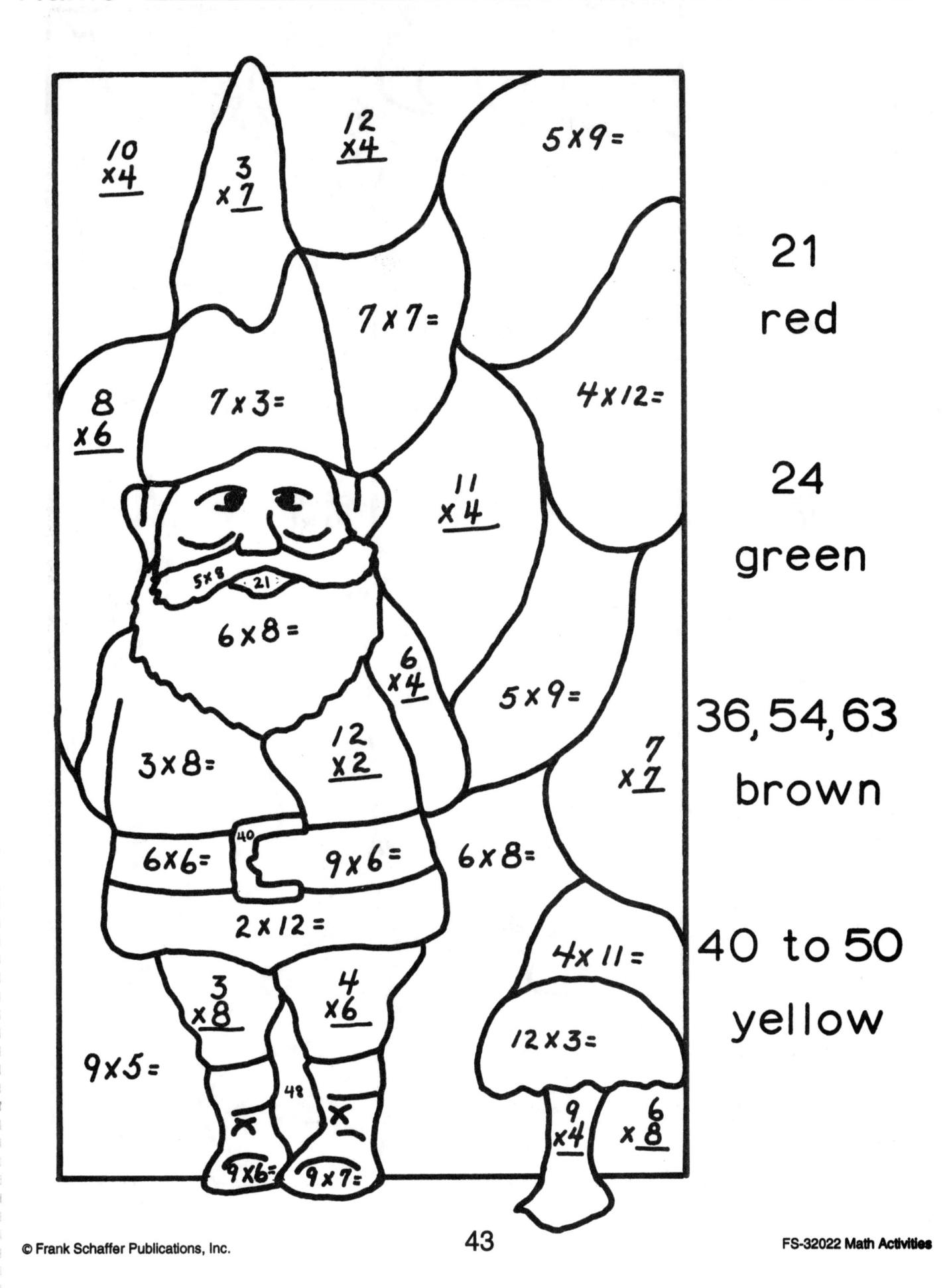

Name _____

$\begin{array}{r}10\\\times 4\\\hline\end{array}$

$\begin{array}{r}3\\\times 7\\\hline\end{array}$

$\begin{array}{r}12\\\times 4\\\hline\end{array}$

5 × 9 =

7 × 7 =

$\begin{array}{r}8\\\times 6\\\hline\end{array}$

7 × 3 =

4 × 12 =

$\begin{array}{r}11\\\times 4\\\hline\end{array}$

5 × 8 · 21

6 × 8 =

$\begin{array}{r}6\\\times 4\\\hline\end{array}$

5 × 9 =

3 × 8 =

$\begin{array}{r}12\\\times 2\\\hline\end{array}$

$\begin{array}{r}7\\\times 7\\\hline\end{array}$

6 × 6 = 40

9 × 6 =

6 × 8 =

2 × 12 =

$\begin{array}{r}3\\\times 8\\\hline\end{array}$

$\begin{array}{r}4\\\times 6\\\hline\end{array}$

4 × 11 =

12 × 3 =

9 × 5 =

48

9 × 6 =

9 × 7 =

$\begin{array}{r}9\\\times 4\\\hline\end{array}$

$\begin{array}{r}6\\\times 8\\\hline\end{array}$

21 red

24 green

36, 54, 63 brown

40 to 50 yellow

43

© Frank Schaffer Publications, Inc.

FS-32022 Math Activities

Name _____

12	20	45, 48, 56	60 to 70
green	yellow	blue	red

Name _____

18	36	56, 63, 64	70 to 85
blue	green	yellow	orange

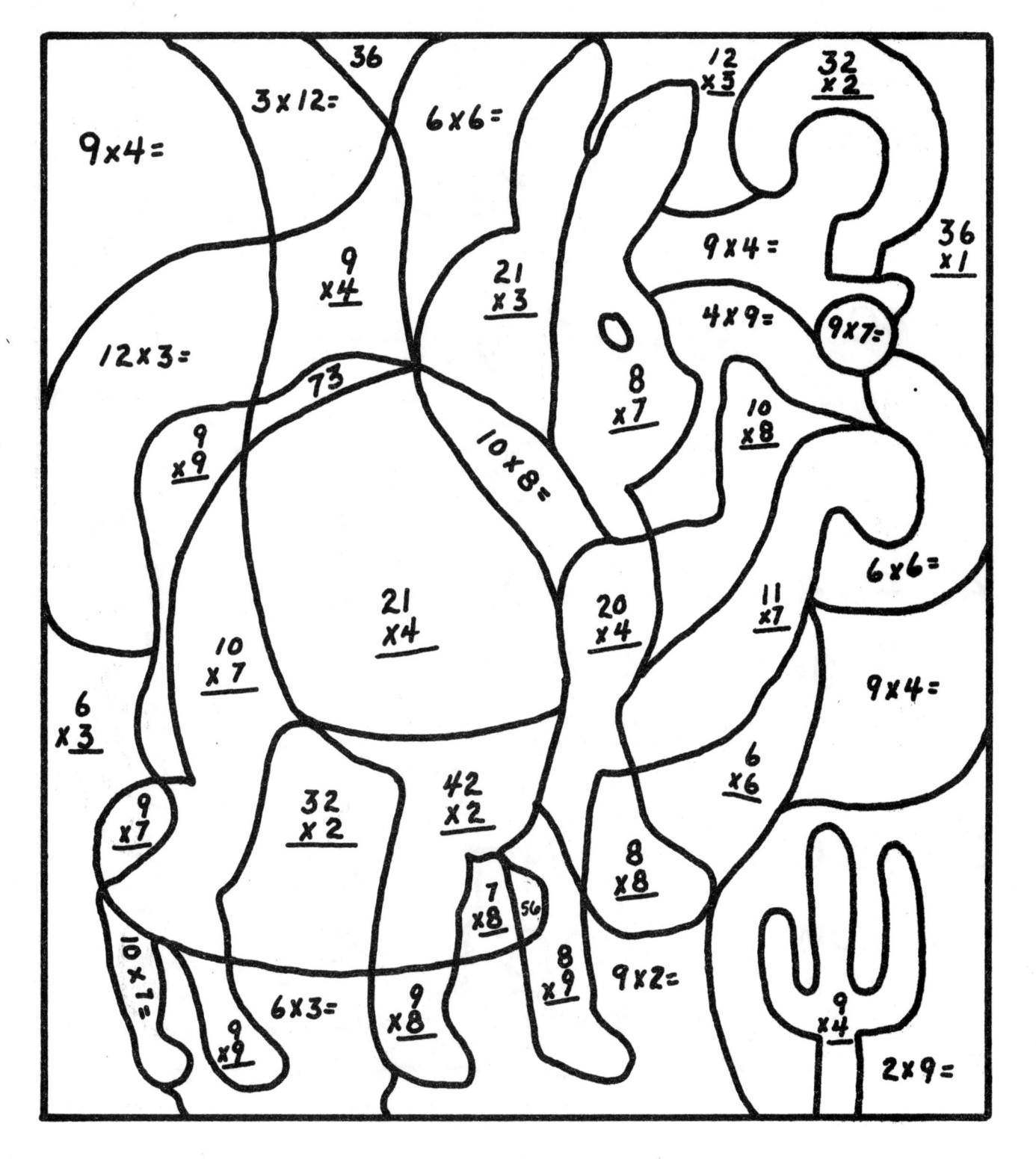

45

Name _____

28 red

40 blue

42, 45, 48 orange

50 to 65 yellow

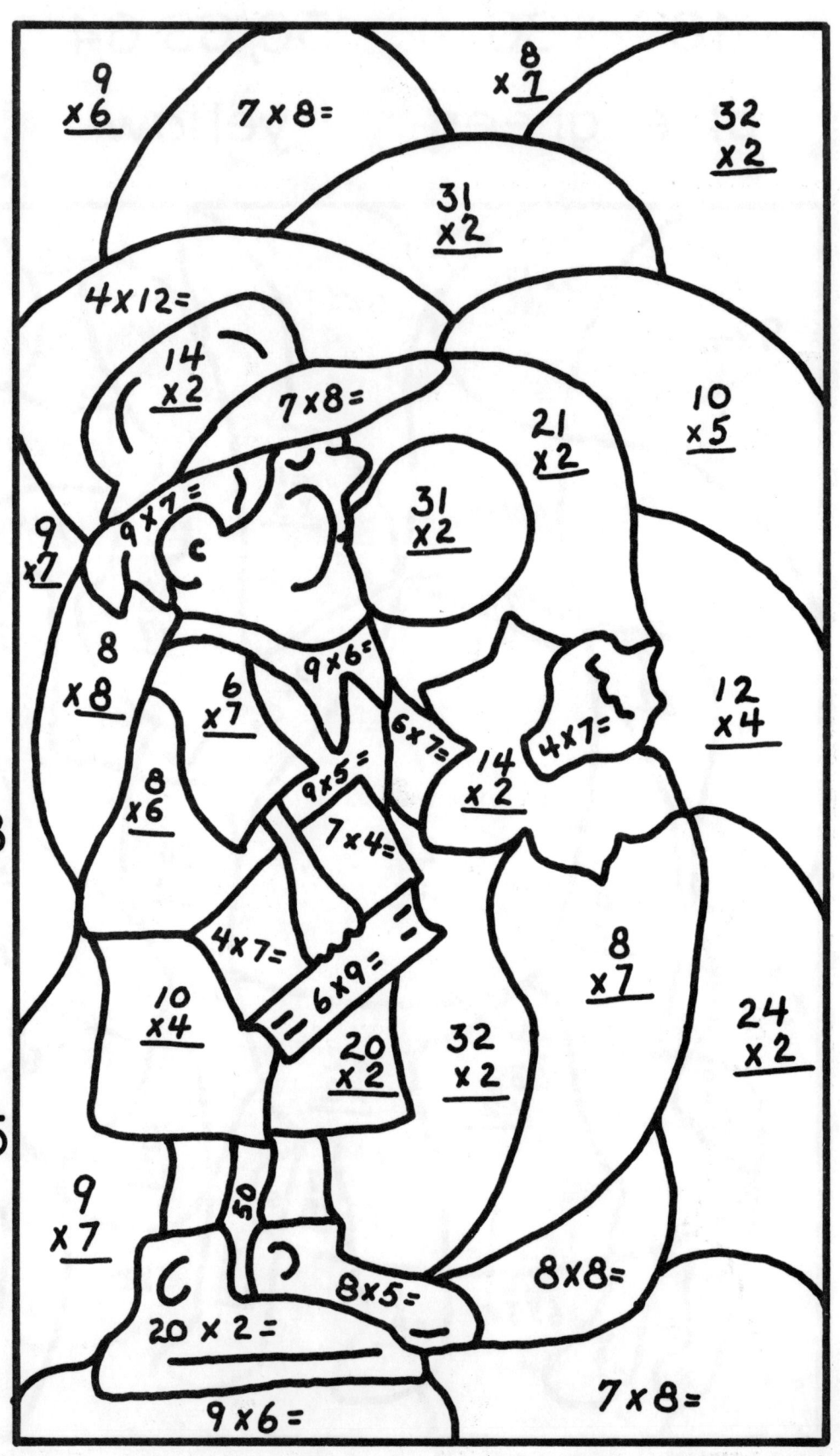

Name _____

28	32	40	56, 64, 72
black	red	orange	yellow

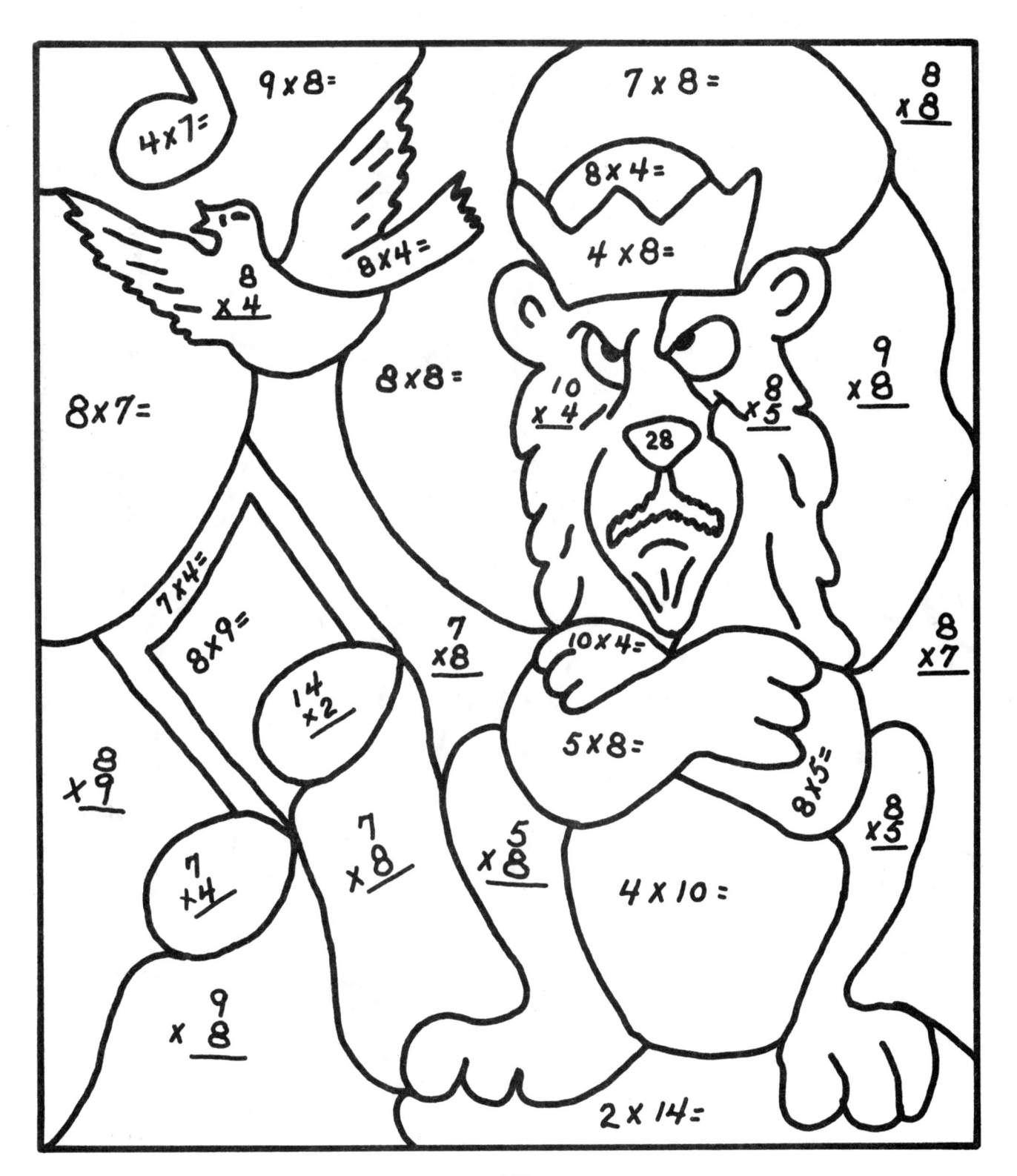

Name _____

18	24	63, 64, 66	70 to 85
yellow	red	orange	black

- 6 × 3
- 4 × 20 =
- 9 × 2
- 7 × 9 =
- 2 × 9 =
- 3 × 6
- 40 × 2
- 9 × 7 =
- 8 × 8
- 22 × 3
- 8 × 8
- 21 × 4
- 8 × 9 =
- 11 × 7
- 24
- 7 × 12 =
- 32 × 2
- 41 × 2
- 10 × 8
- 9 × 9 =
- 33 × 2
- 63
- 7 × 10 =
- 11 × 7 =
- 6 × 4
- 3 × 8 =
- 8 × 3
- 80
- 11 × 6
- 12 × 2
- 8 × 3 =
- 2 × 12 =
- 6 × 4 =
- 8 × 8 =
- 7 × 9 =
- 6 × 3 =

48

© Frank Schaffer Publications, Inc.   FS-32022 Math Activities

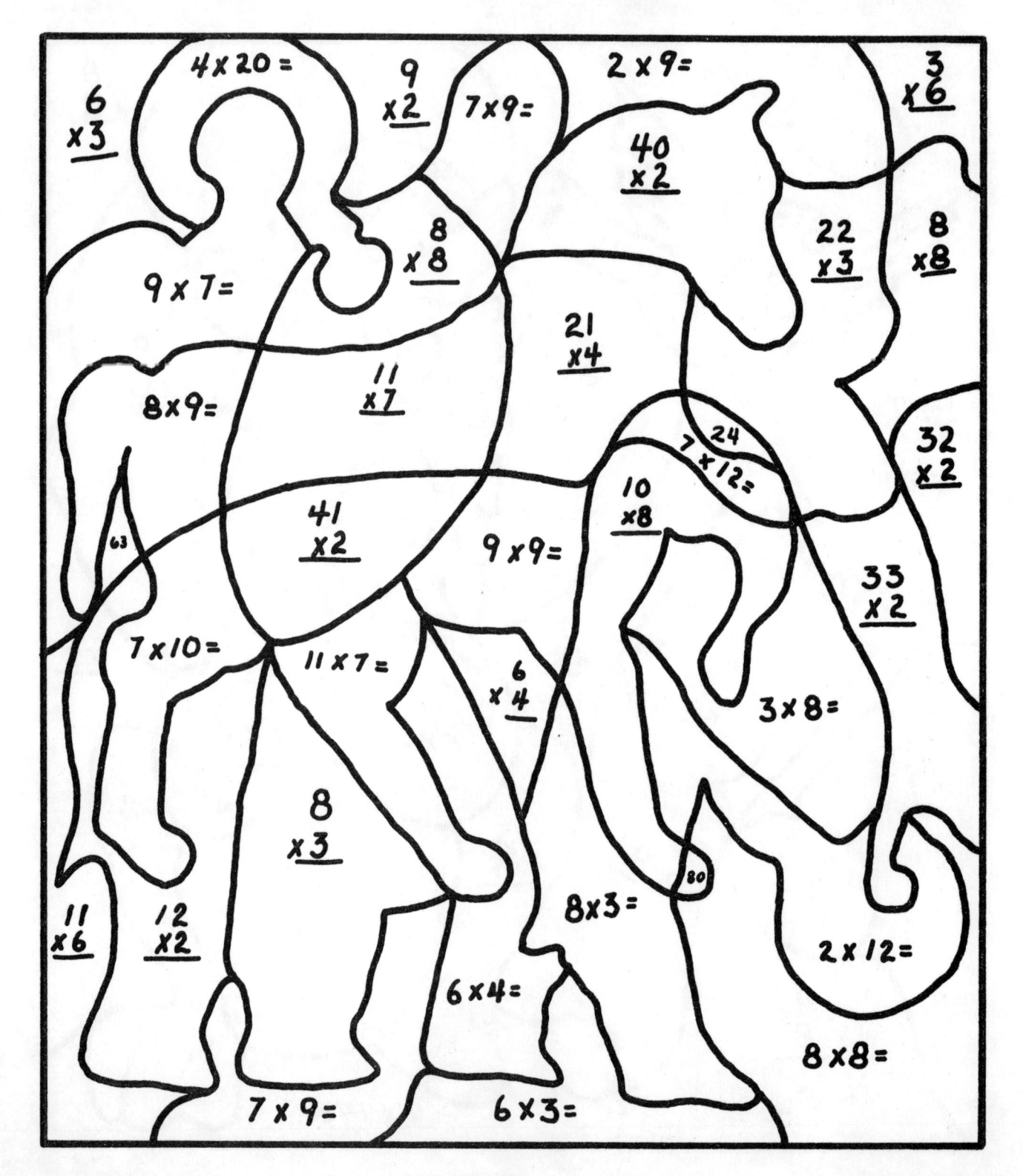

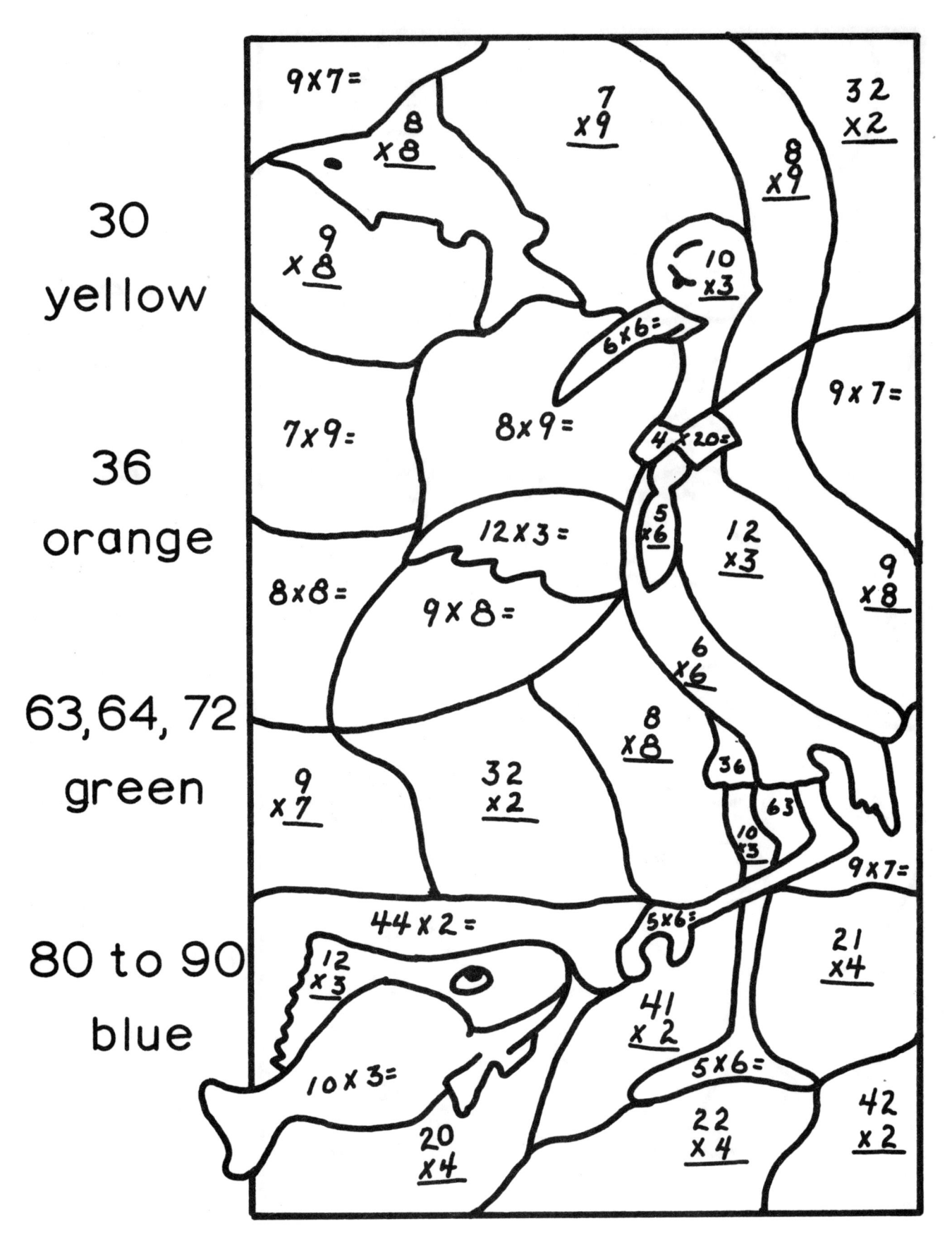

Name _____

30 yellow

36 orange

63, 64, 72 green

80 to 90 blue

9×7=   8 ×8   7 ×9   8 ×9   32 ×2
9 ×8   10 ×3
                          6×6=
7×9=   8×9=   4×20=   9×7=
8×8=   12×3=   5 ×6   12 ×3   9 ×8
       9×8=   6 ×6
              8 ×8   36
9 ×7   32 ×2   10 ×3   63
                              9×7=
44×2=   5×6=   21 ×4
12 ×3   41 ×2
10×3=   20 ×4   5×6=   22 ×4   42 ×2

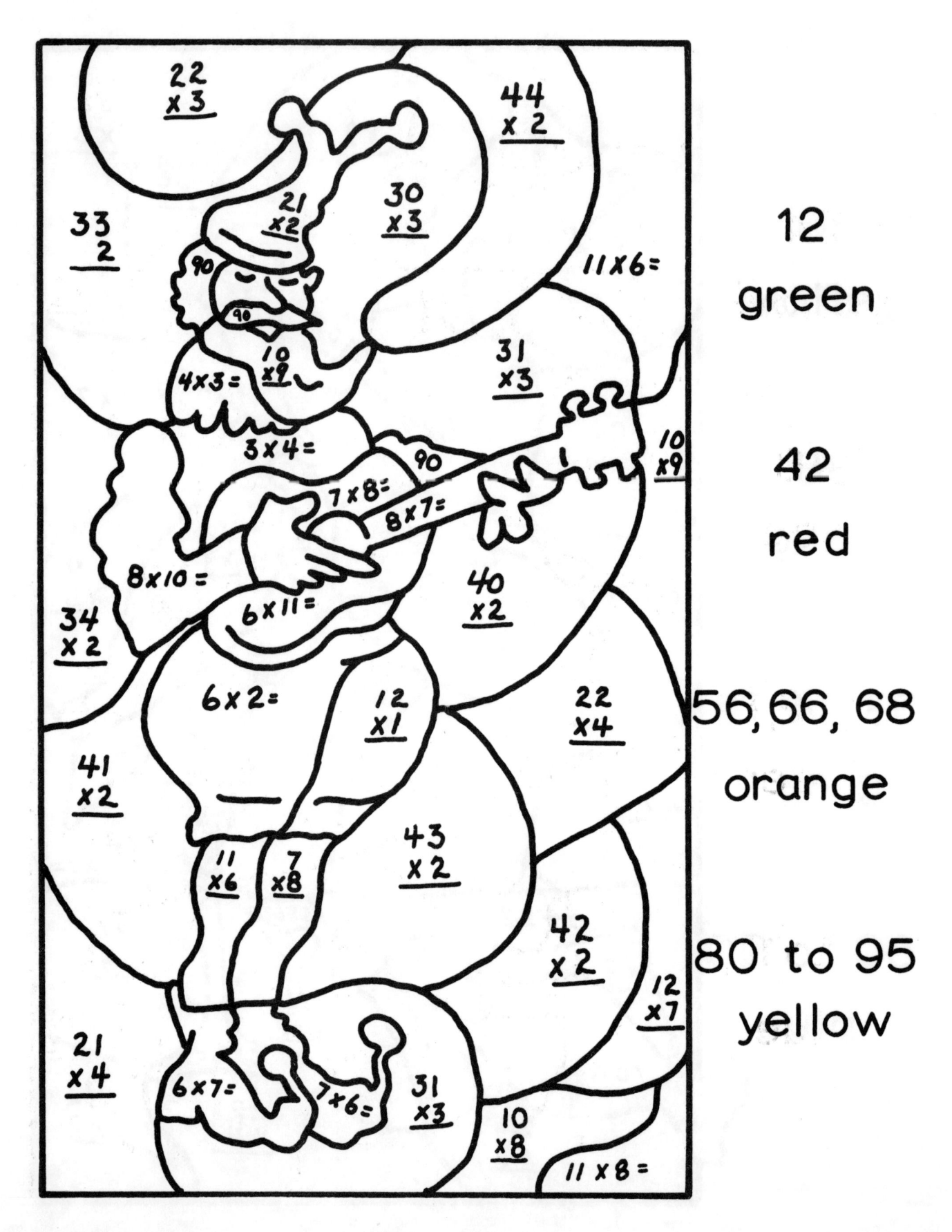

Name _____

12 green

42 red

56, 66, 68 orange

80 to 95 yellow

50

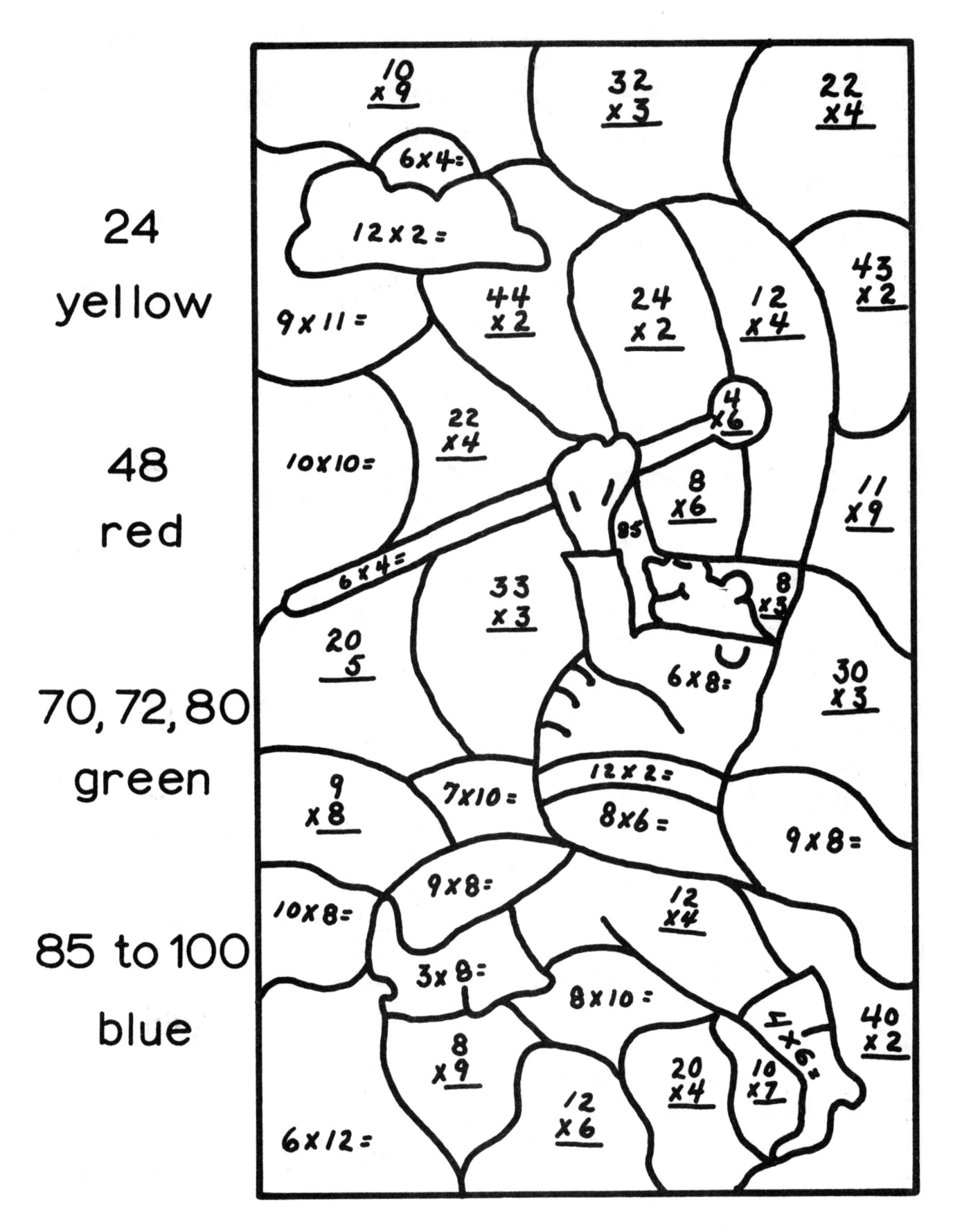

Name _____

24
yellow

48
red

70, 72, 80
green

85 to 100
blue

10 x 9

32 x 3

22 x 4

6 x 4 =

12 x 2 =

44 x 2

24 x 2

12 x 4

43 x 2

9 x 11 =

4 x 6

22 x 4

10 x 10 =

8 x 6

11 x 9

6 x 4 =

33 x 3

85

8 x 3

20 x 5

6 x 8 =

30 x 3

9 x 8

7 x 10 =

12 x 2 =

8 x 6 =

9 x 8 =

10 x 8 =

9 x 8 =

12 x 4

3 x 8 =

8 x 10 =

4 x 6 =

40 x 2

8 x 9

20 x 4

10 x 7

6 x 12 =

12 x 6

© Frank Schaffer Publications, Inc.   51   FS-32022 Math Activities

Name _____

Skill: division by 2

1	2	3	4	5
blue	green	yellow	brown	orange

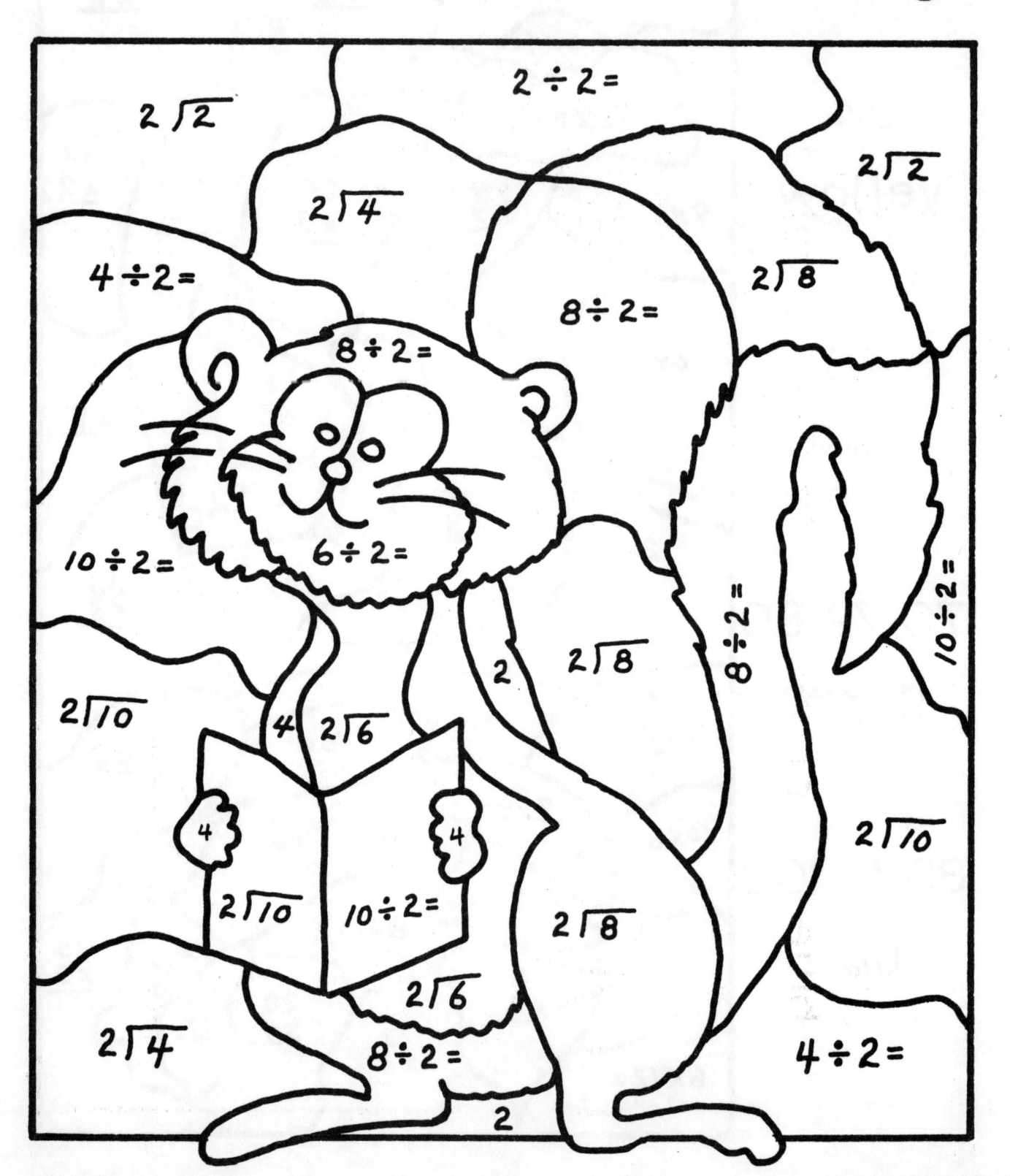

$2 \overline{)2}$

$2 \div 2 =$

$2 \overline{)2}$

$2 \overline{)4}$

$4 \div 2 =$

$8 \div 2 =$

$2 \overline{)8}$

$8 \div 2 =$

$10 \div 2 =$

$6 \div 2 =$

$10 \div 2 =$

2

$2 \overline{)8}$

$8 \div 2 =$

$2 \overline{)10}$

4

$2 \overline{)6}$

4

$2 \overline{)10}$

$2 \overline{)10}$

$10 \div 2 =$

$2 \overline{)8}$

$2 \overline{)6}$

$2 \overline{)4}$

$8 \div 2 =$

$4 \div 2 =$

2

© Frank Schaffer Publications, Inc.

FS-32022 Math Activities

Name _____

Skill: division by 2

6	7	8	9	10
blue	orange	brown	yellow	green

© Frank Schaffer Publications, Inc.

53

FS-32022 Math Activities

Name _____

Skill: division by 3

1	2	3	4	5
black	yellow	green	red	orange

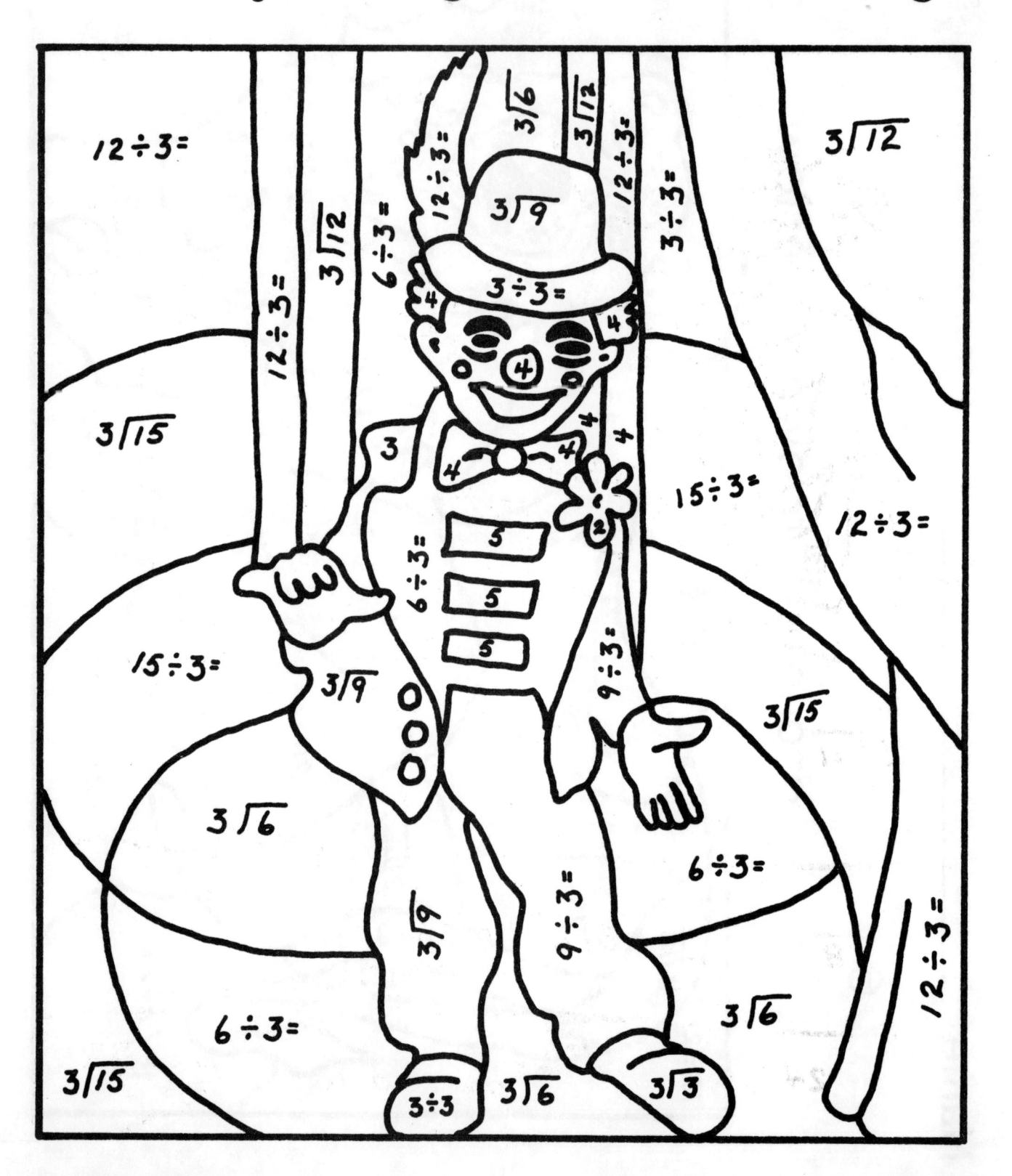

Name _____

Skill: division by 3

6	7	8	9	10
brown	yellow	orange	green	red

$21 \div 3 =$ $3\overline{)21}$ $30 \div 3 =$

$30 \div 3 =$ $21 \div 3 =$ $18 \div 3 =$ $24 \div 3 =$ 7 $21 \div 3 =$

$3\overline{)21}$ $3\overline{)21}$ $3\overline{)30}$

$3\overline{)27}$ $27 \div 3 =$

$3\overline{)27}$

$30 \div 3 =$

$3\overline{)21}$ $21 \div 3 =$ $3\overline{)21}$

$27 \div 3 =$ $3\overline{)27}$

$3\overline{)18}$ $3\overline{)24}$ $3\overline{)18}$ $18 \div 3 =$

$18 \div 3 =$

$24 \div 3 =$ $3\overline{)18}$ $3\overline{)24}$

© Frank Schaffer Publications, Inc.

Skill: division by 4

Name _____

1	2	3	4	5
blue	black	yellow	red	orange

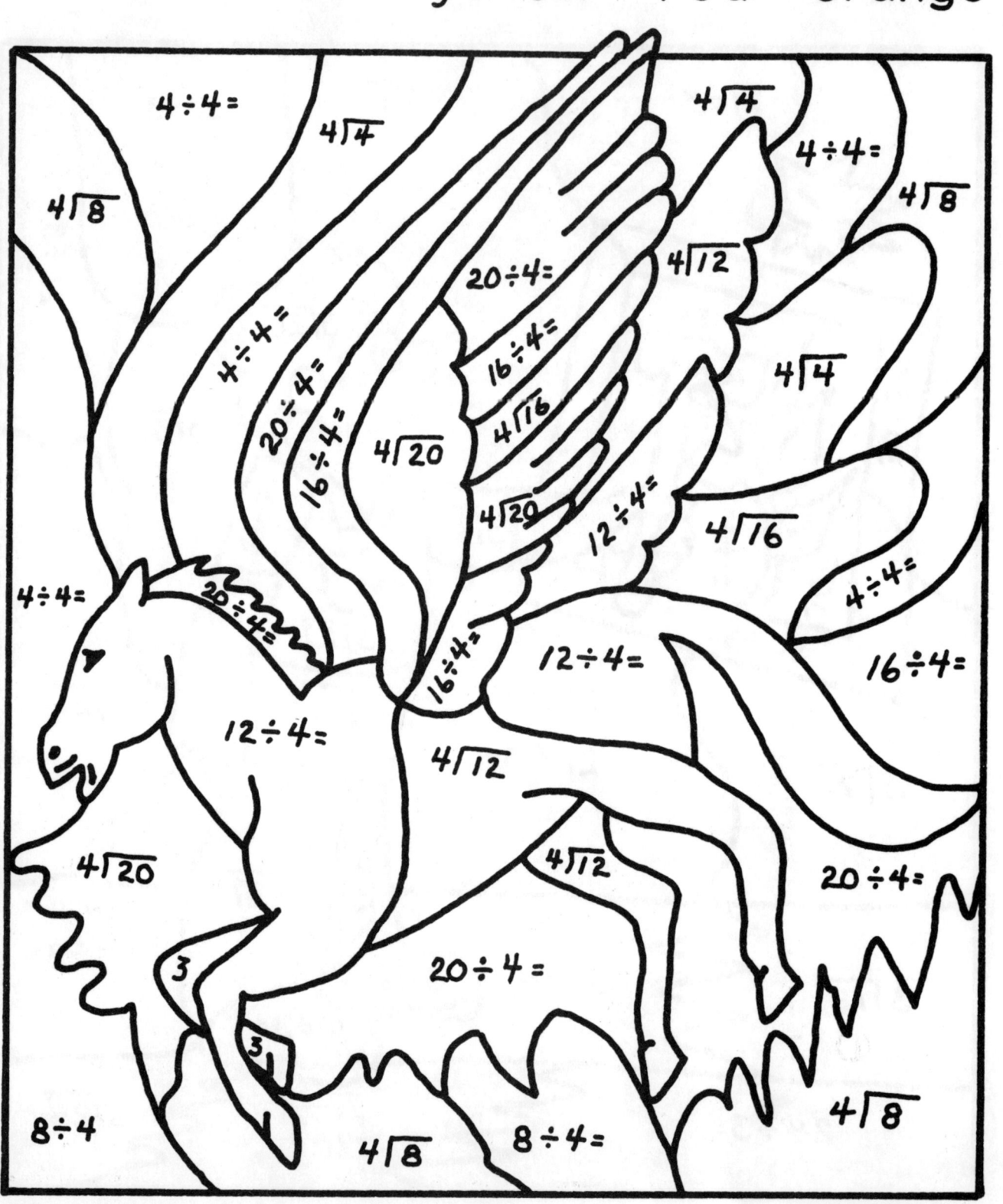

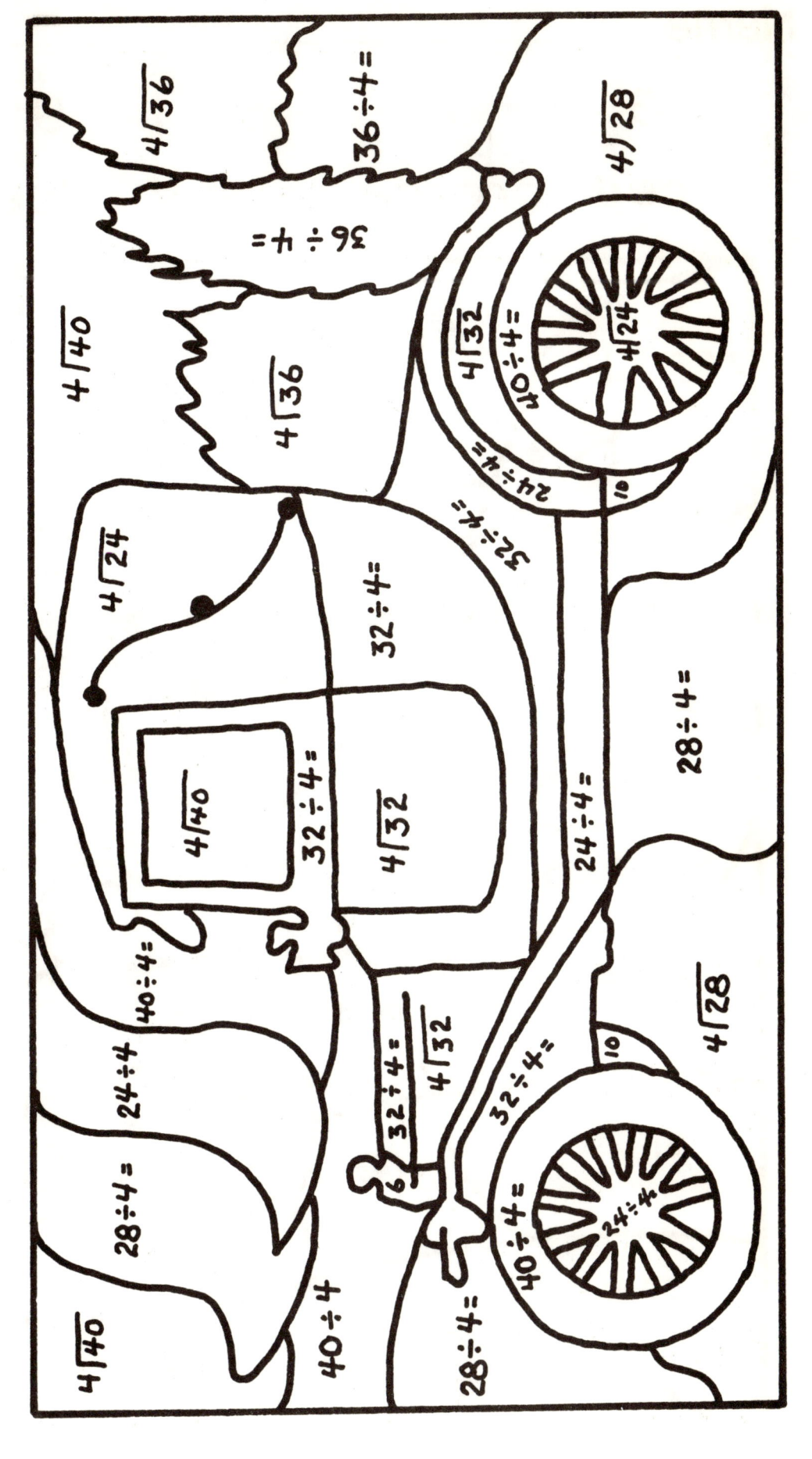

Name _____

Skill: division by 4

6	7	8	9	10
yellow	orange	red	green	blue

Name _____

Skill: division by 5

6	7	8	9	10
red	orange	yellow	blue	brown

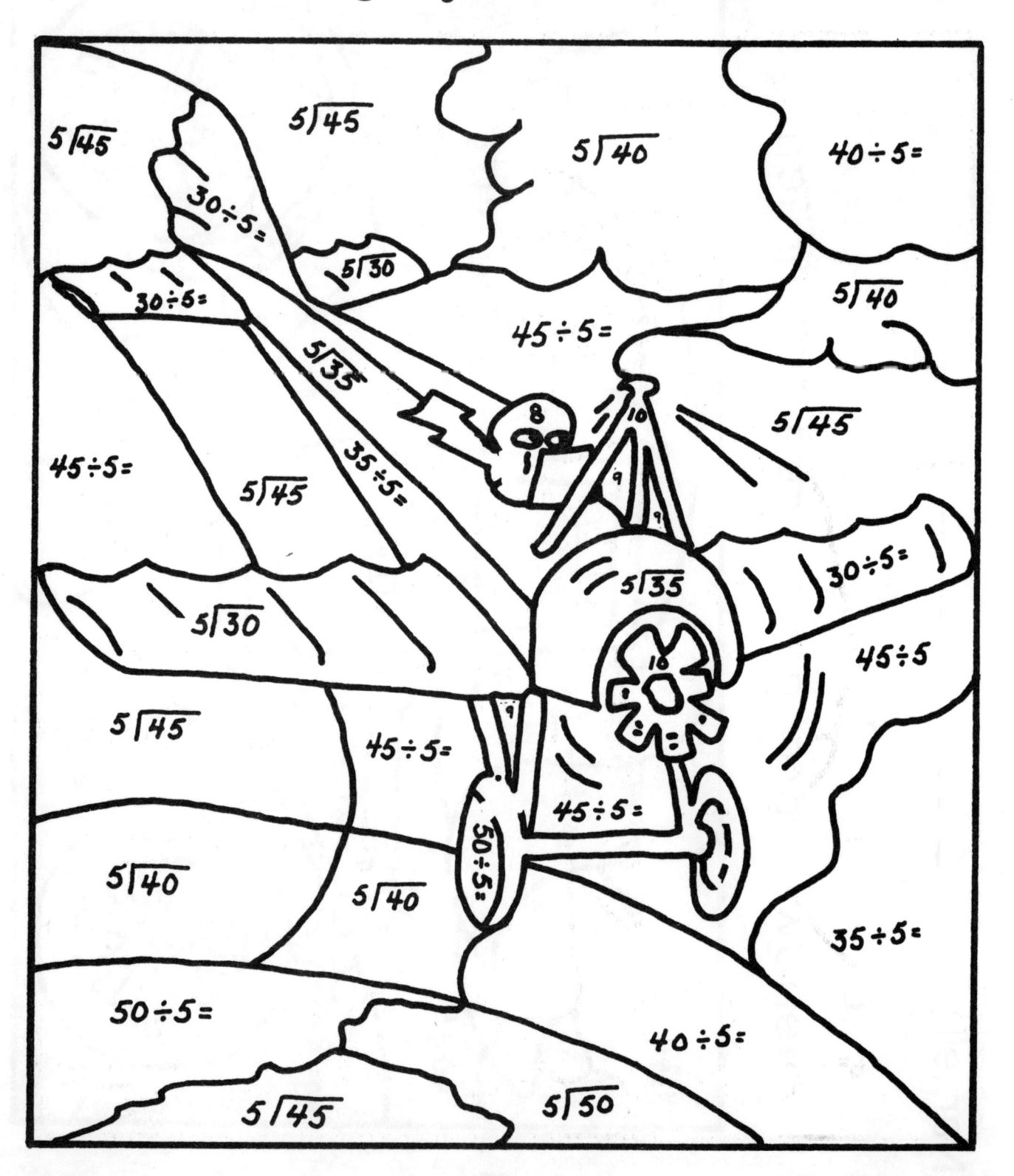

Name _____

Skill: division by 6

6	7	8	9	10
blue	yellow	brown	red	green

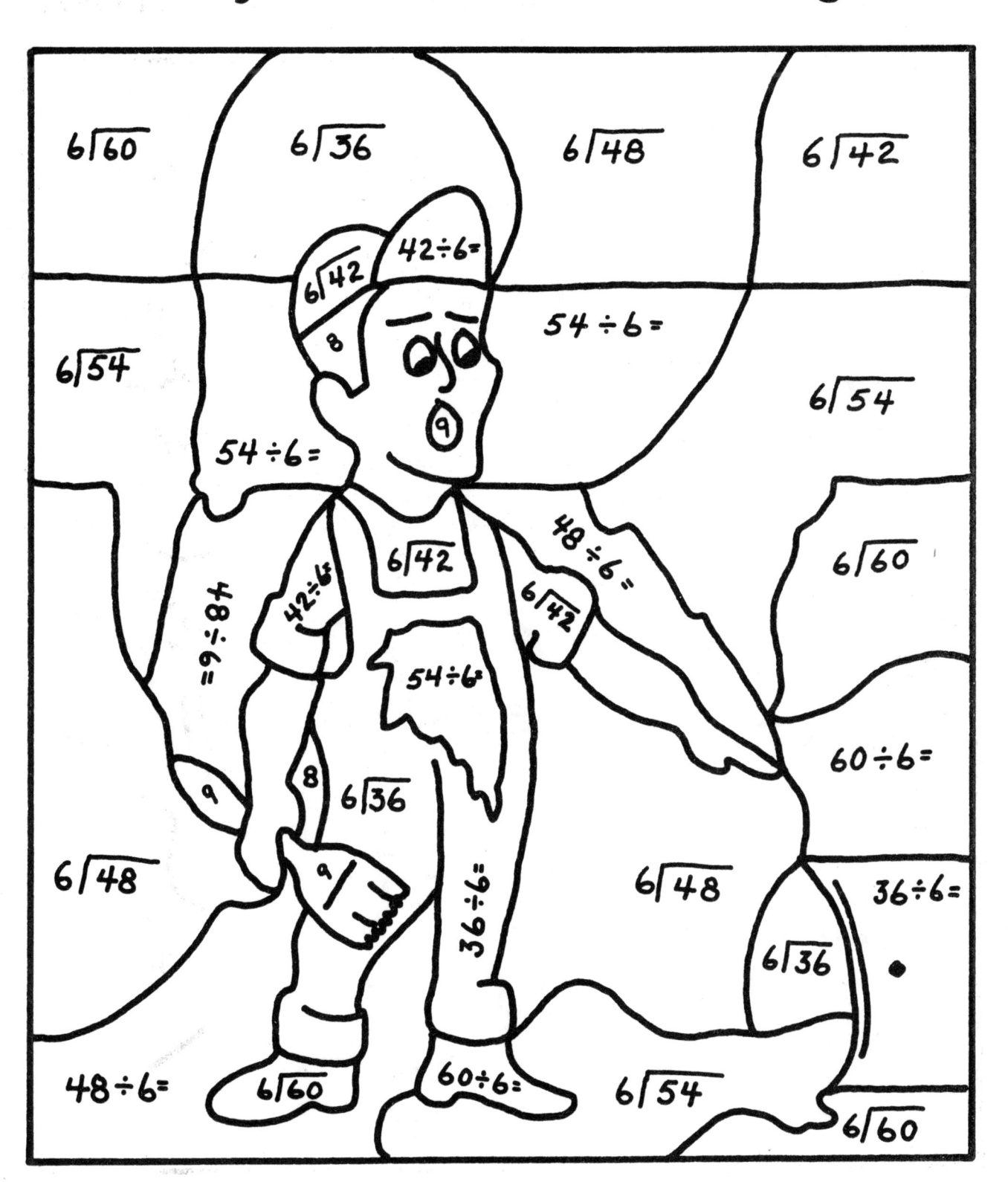

Name _____

Skill: division by 6

1	2	3	4	5
red	yellow	brown	green	orange

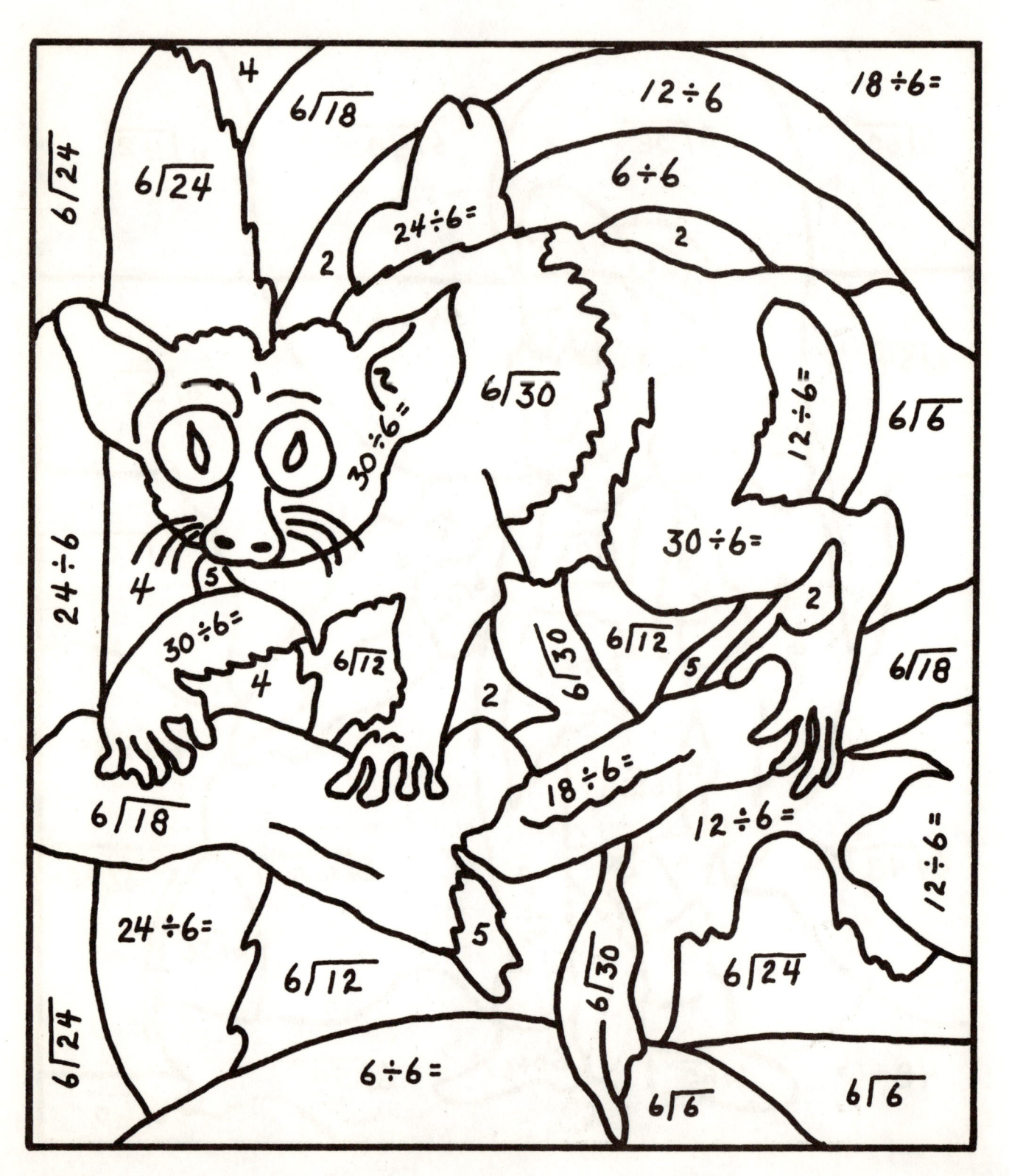

Skill: division by 7

Name _____

1	2	3	4	5
blue	yellow	orange	brown	green

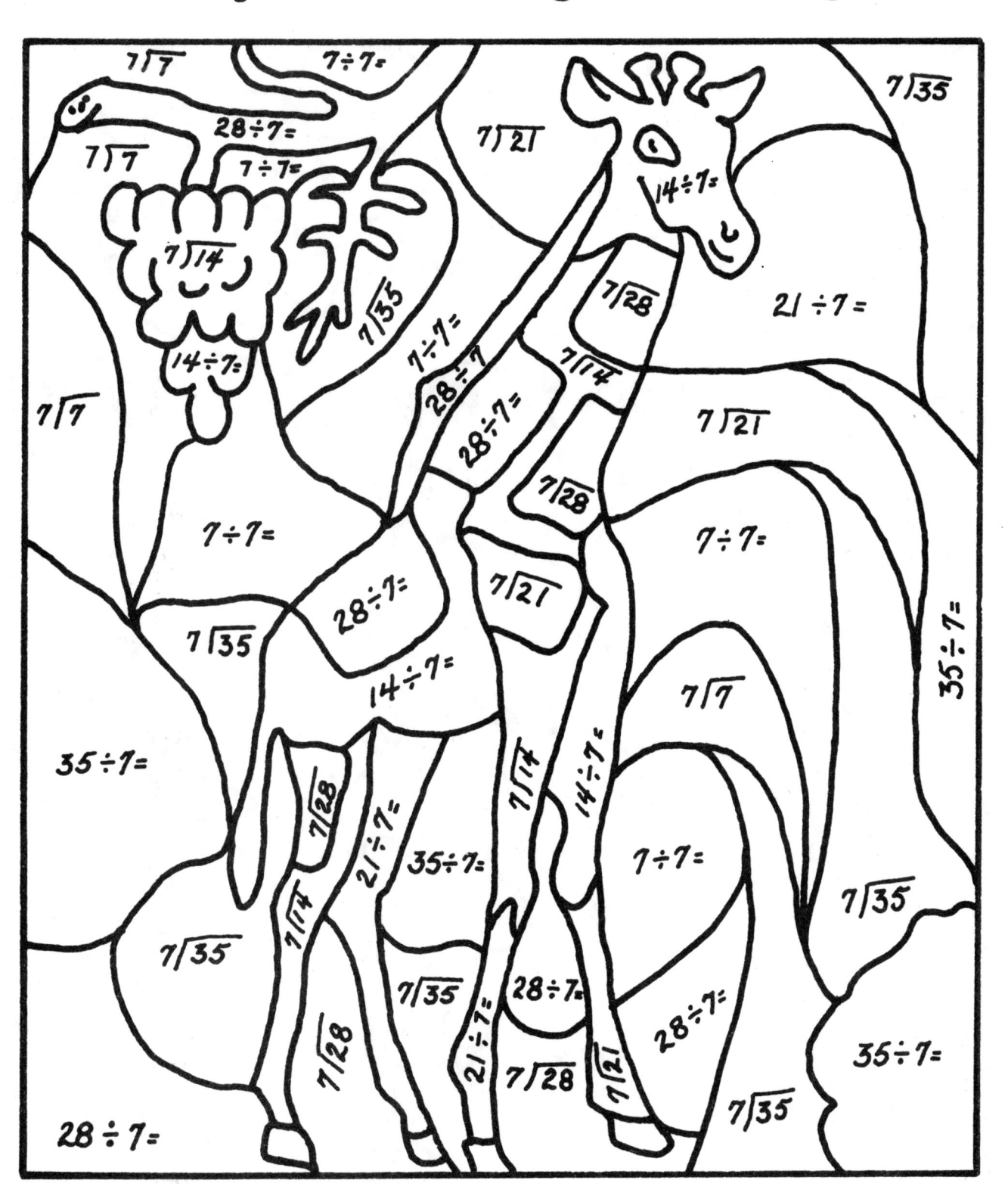

Name _____

Skill: division by 7

    6   7   8   9   10

red   brown orange yellow blue

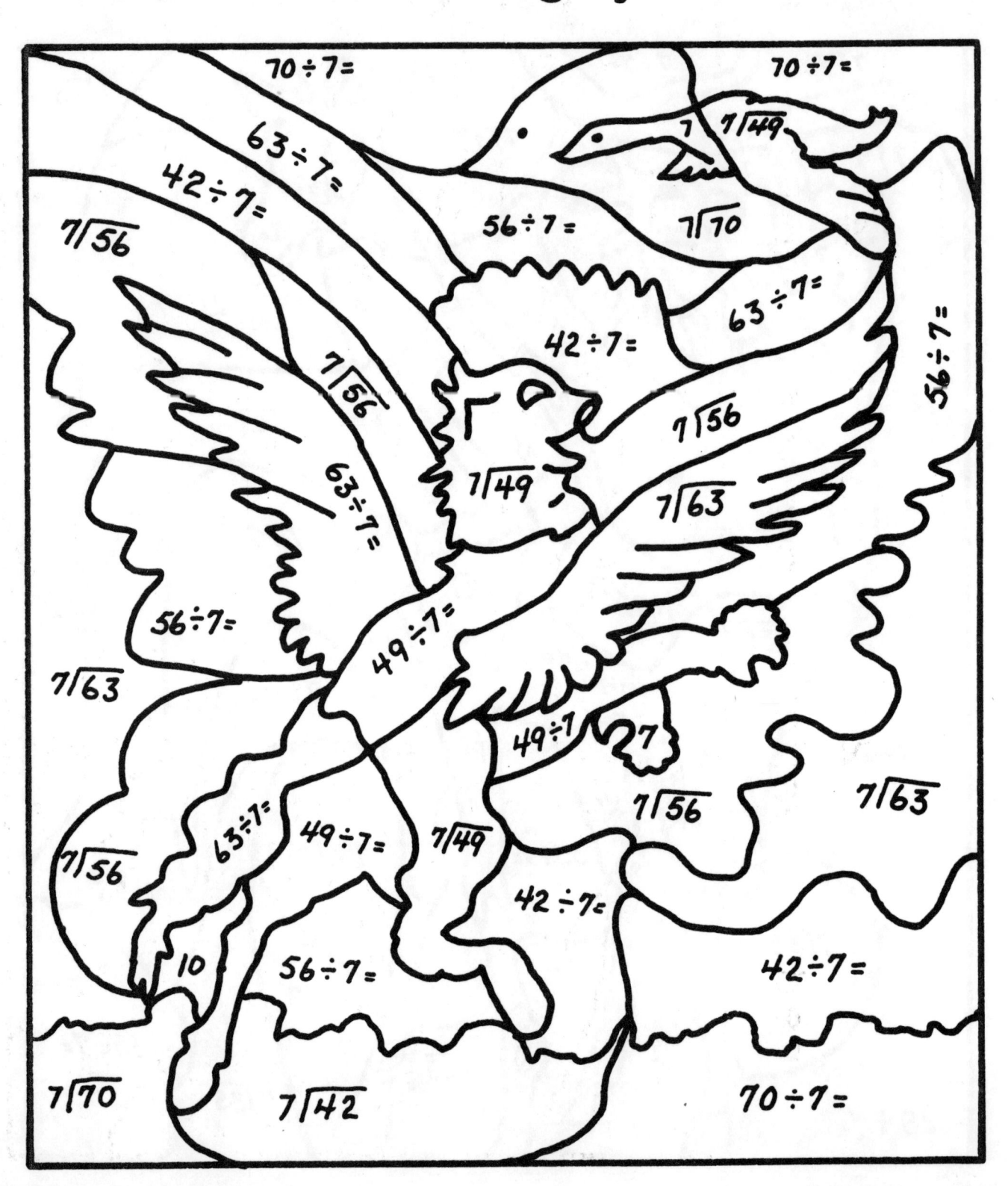

© Frank Schaffer Publications, Inc.   62   FS-32022 Math Activities

Name _____

Skill: division by 8

1	2	3	4	5
red	yellow	orange	purple	blue

Name _____

Skill: division by 8

6	7	8	9	10
brown	yellow	green	orange	red

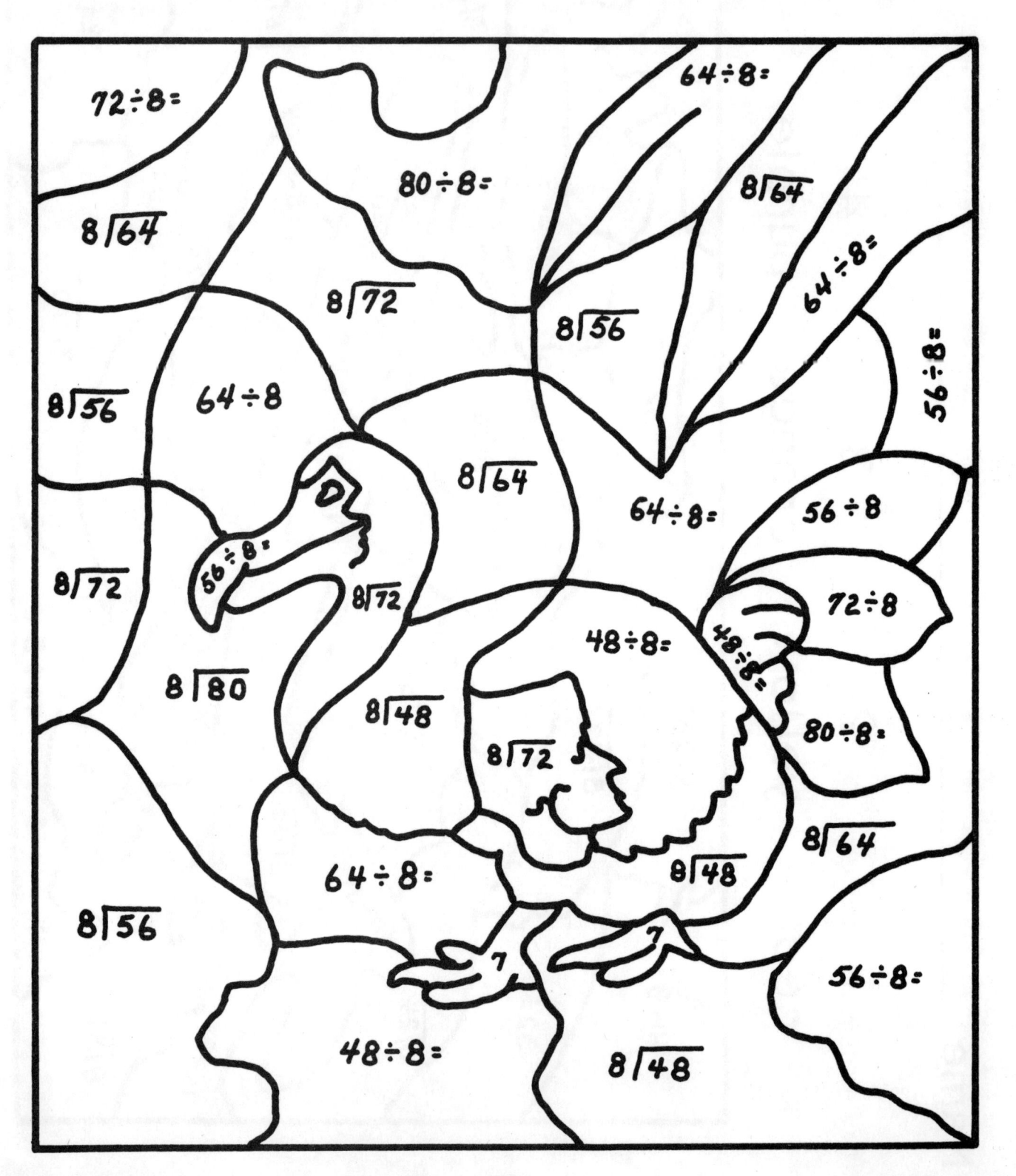

$72 \div 8 =$

$8\overline{)64}$

$80 \div 8 =$

$8\overline{)72}$

$64 \div 8 =$

$8\overline{)64}$

$64 \div 8 =$

$8\overline{)56}$

$8\overline{)56}$

$64 \div 8$

$8\overline{)64}$

$64 \div 8 =$

$56 \div 8 =$

$56 \div 8$

$8\overline{)72}$

$56 \div 8 =$

$8\overline{)72}$

$72 \div 8$

$8\overline{)80}$

$48 \div 8 =$

$48 \div 8 =$

$8\overline{)48}$

$80 \div 8 =$

$8\overline{)72}$

$8\overline{)64}$

$64 \div 8 =$

$8\overline{)48}$

$8\overline{)56}$

$56 \div 8 =$

$48 \div 8 =$

$8\overline{)48}$

© Frank Schaffer Publications, Inc.

64

FS-32022 Math Activities

Name _____

Skill: division by 9

1	2	3	4	5
red	green	yellow	brown	orange

Name _____

Skill: division by 9

6	7	8	9	10
blue	yellow	brown	red	orange

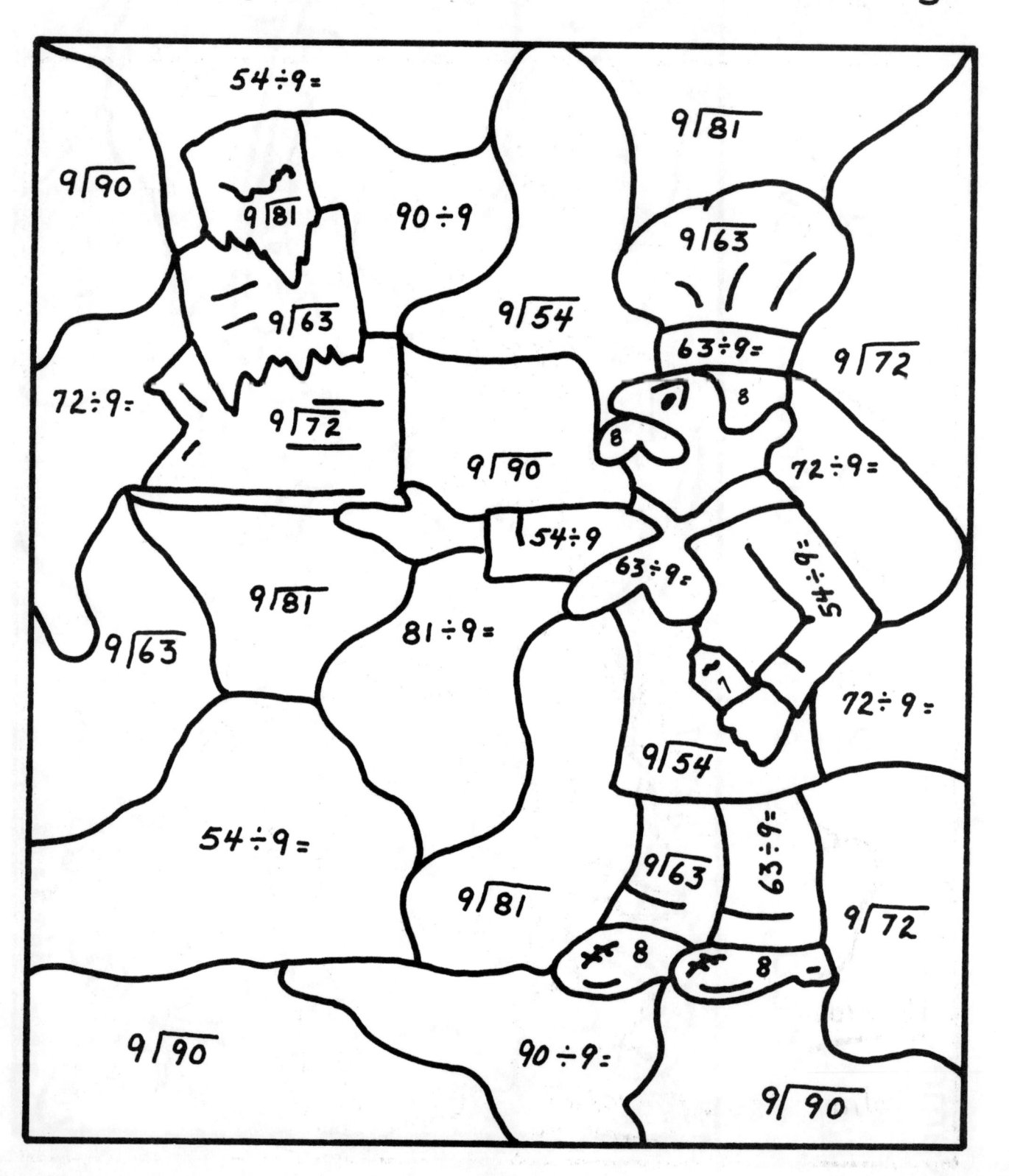

$54 \div 9 =$
$9\overline{)90}$
$9\overline{)81}$
$90 \div 9$
$9\overline{)81}$
$9\overline{)63}$
$9\overline{)54}$
$9\overline{)63}$
$72 \div 9 =$
$9\overline{)72}$
$9\overline{)72}$
$9\overline{)90}$
$54 \div 9$
$63 \div 9 =$
$63 \div 9 =$
$54 \div 9 =$
$72 \div 9 =$
$9\overline{)81}$
$9\overline{)63}$
$81 \div 9 =$
$72 \div 9 =$
$9\overline{)54}$
$54 \div 9 =$
$9\overline{)63}$
$63 \div 9 =$
$9\overline{)81}$
$9\overline{)72}$
$9\overline{)90}$
$90 \div 9 =$
$9\overline{)90}$

© Frank Schaffer Publications, Inc. FS-32022 Math Activities

Name _____

Skill: division by 10

1	2	3	4	5
orange	red	blue	yellow	brown

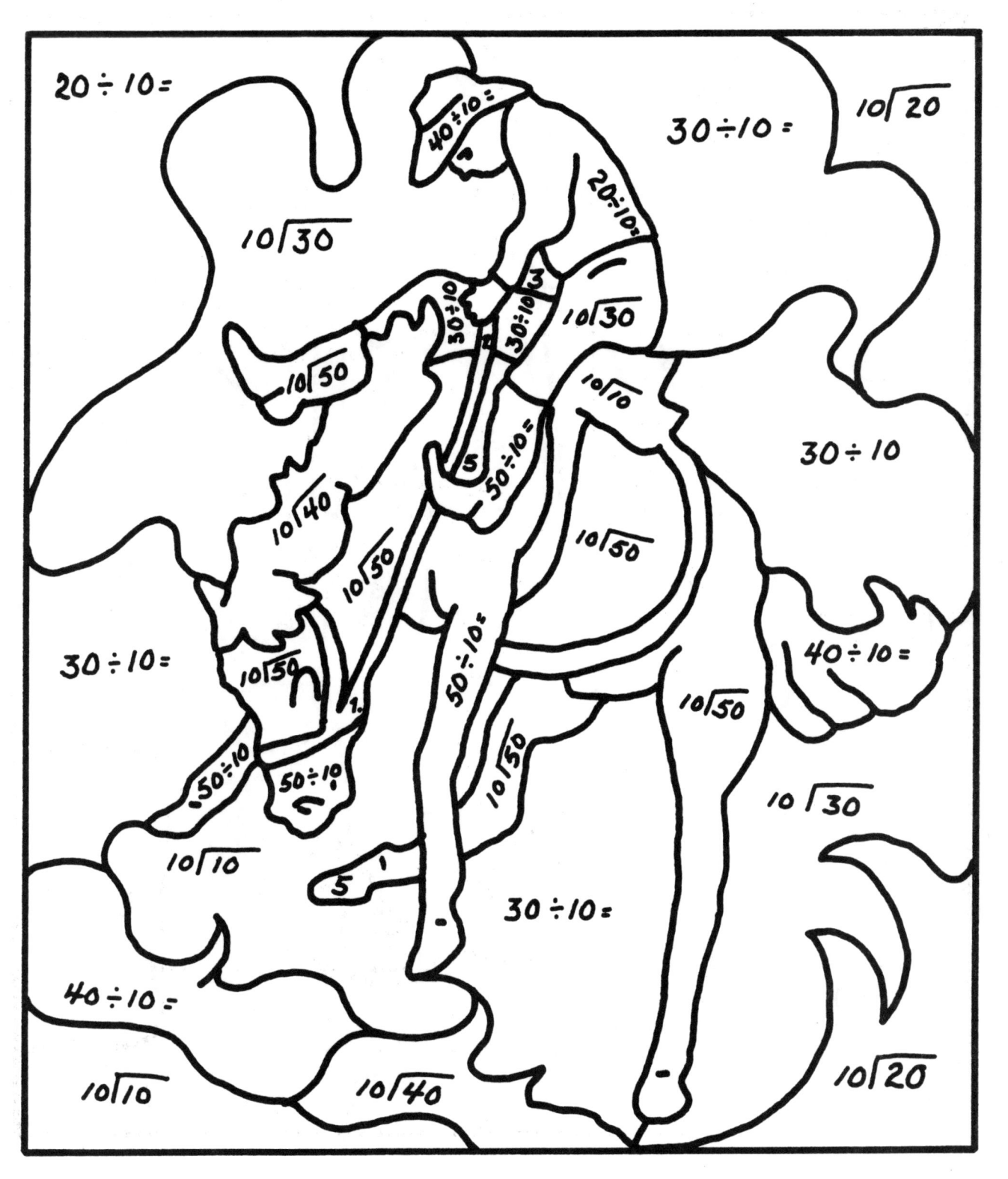

© Frank Schaffer Publications, Inc.

67

FS-32022 Math Activities

Name _____

Skill: division by 10

6	7	8	9	10
blue	yellow	red	brown	green

Name _____   Skill: Recognizing equal parts

A fraction is one or more **equal** parts of a whole.

This circle is divided into 2 equal parts.

This circle is not divided into 2 equal parts.

The following figures are divided into how many equal parts?

Put a check (✓) next to the figures **not** divided into equal parts.

LOST PARTS

Name _____ Skill: Writing fractions

A fraction is one or more equal parts of a whole.

3 equal parts
thirds

5 equal parts
fifths

Write the word below its matching figure.

Halves
Thirds
Fourths
Fifths
Sixths
Sevenths
Eighths
Ninths
Tenths

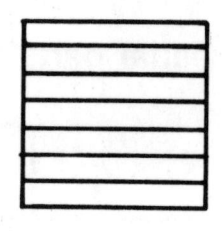

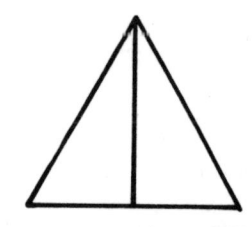

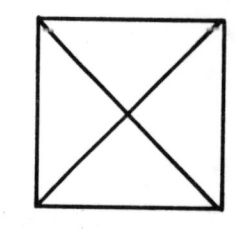

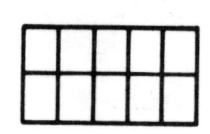

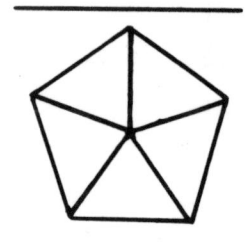

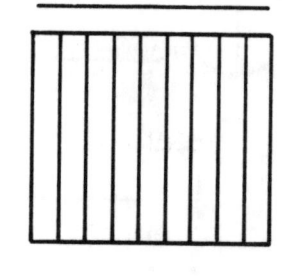

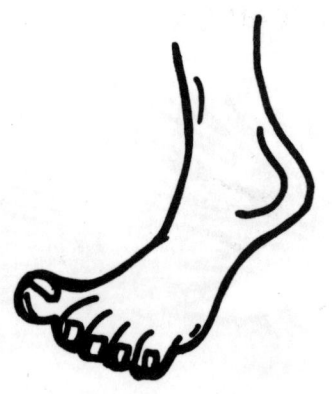

© Frank Schaffer Publications, Inc.

70

FS-32022 Math Activities

Name _____   Skill: Naming one part

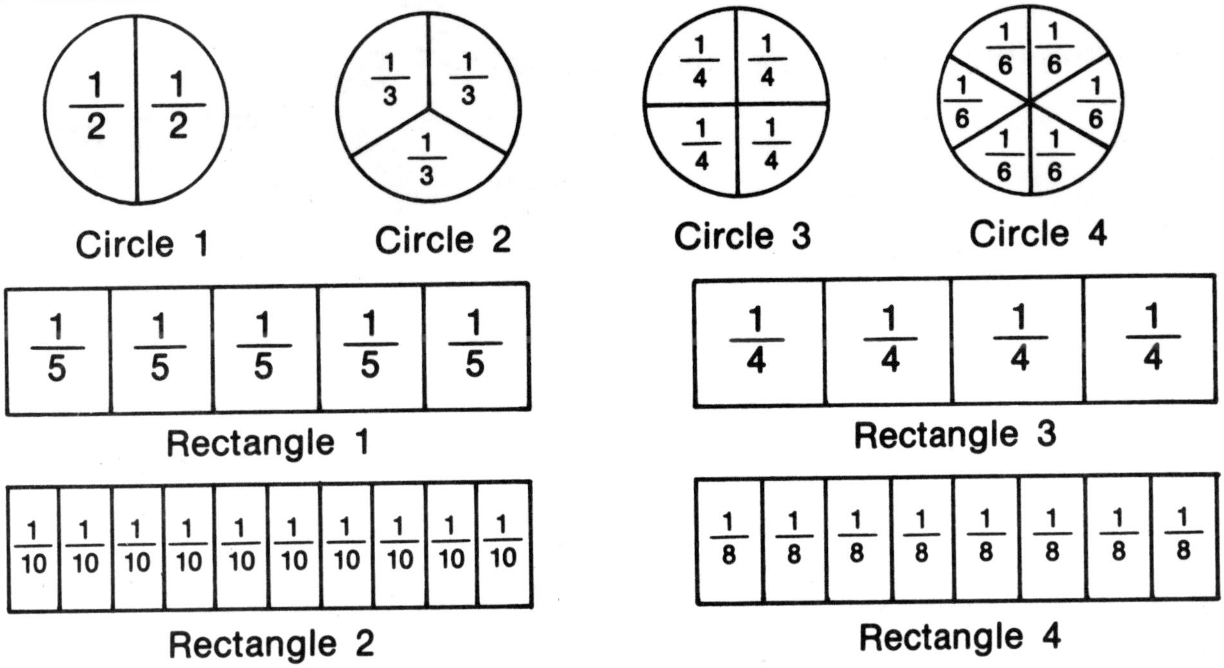

Circle 1 has ____ parts.   The name for each part is ____ .

Circle 2 has ____ parts.   The name for each part is ____ .

Circle 3 has ____ parts.   The name for each part is ____ .

Circle 4 has ____ parts.   The name for each part is ____ .

Rectangle 1 has ____ parts.   The name for each part is ____ .

Rectangle 2 has ____ parts.   The name for each part is ____ .

Rectangle 3 has ____ parts.   The name for each part is ____ .

Rectangle 4 has ____ parts.   The name for each part is ____ .

Name _____   Skill: **Writing fractions**

Write fractions for:

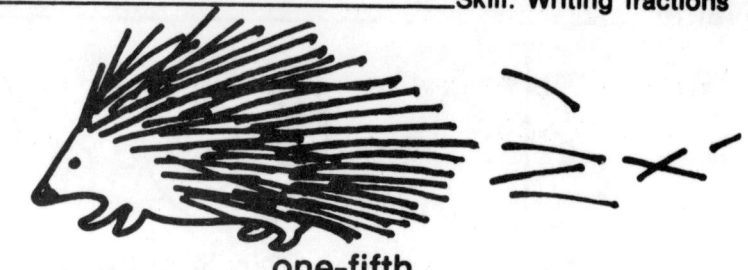

one-half _____   one-third _____   one-fifth _____

one-fourth _____   one-sixth _____   one-tenth _____

one-eighth _____   one-twelfth _____   one-seventh _____

Write fractions for:

one of four parts _____   one of nine parts _____

one of three parts _____   one of six parts _____

one of two parts _____   one of five parts _____

Color each part as directed.

Color $\frac{1}{3}$ blue. 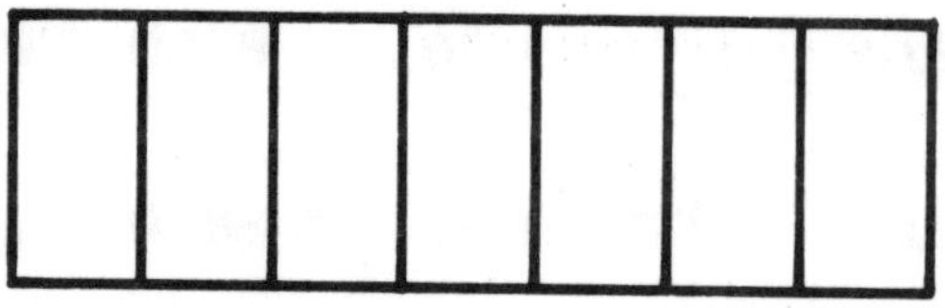   Color $\frac{1}{7}$ yellow. 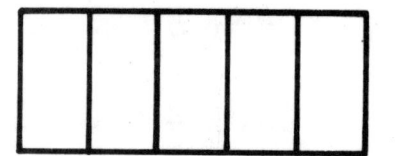   Color $\frac{1}{5}$ red.

Color $\frac{1}{4}$ green. 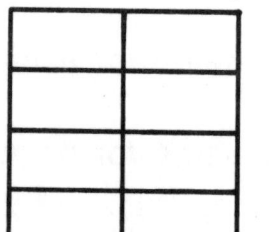   Color $\frac{1}{8}$ orange. 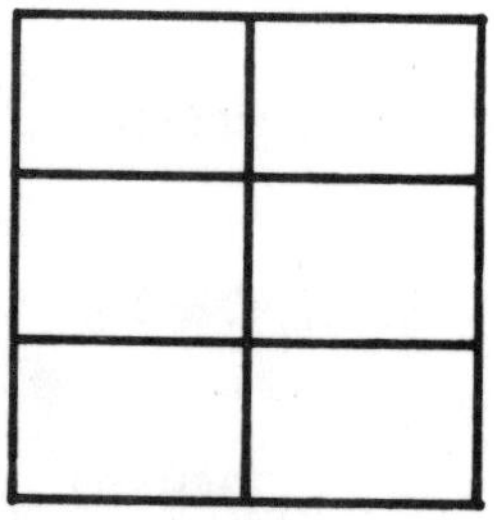   Color $\frac{1}{6}$ brown. 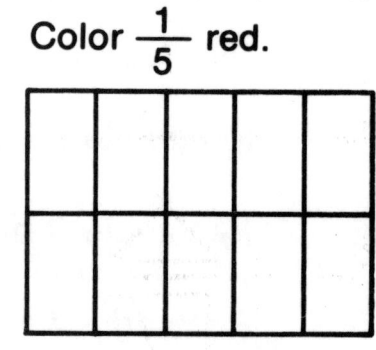   Color $\frac{1}{10}$ black.

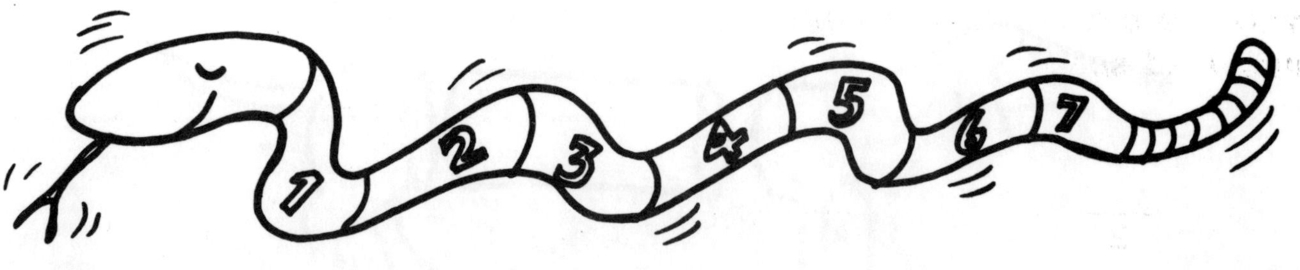

72

© Frank Schaffer Publications, Inc.   FS-32022 Math Activities

Name _____ Skill: Equal parts

What is the easiest way to make a banana split?

Look at each figure below. If it is divided into equal parts, circle the letter under YES. If it is not divided into equal parts, circle the letter under NO.

#	Figure	YES	NO
1.	triangle	B	C
2.	square with X	U	A
3.	circle in 4	T	C
4.	triangle divided	I	K
5.	rectangle	R	T

#	Figure	YES	NO
6.	circle	O	I
7.	triangle	N	M
8.	rectangle in parts	H	S
9.	square	R	A
10.	circle	I	L
11.	square with diamond	E	F

Write the circled letters in the numbered spaces below.

___ ___ ___   ___ ___   ___ ___   ___ ___ ___ ___
 1   2   3   4   5   6   7   8   9 10 11

73

Name _____ Skill: Writing and shading fractions

Here are two of three equal parts of this circle.

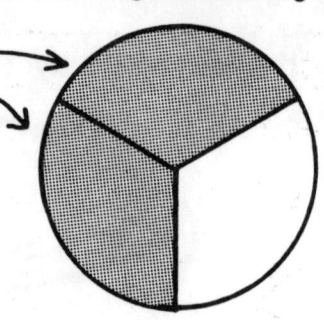

The fraction for "two of three parts" is $\frac{2}{3}$.

Fill in the missing numbers and shade the parts.

"five of six" = $\frac{}{6}$

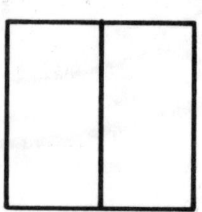

"four of five" = $\frac{4}{}$

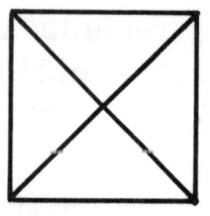

"three of eight" = $\frac{}{8}$

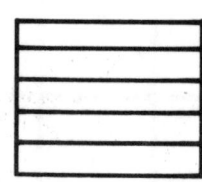

"three of four" = $\frac{3}{}$

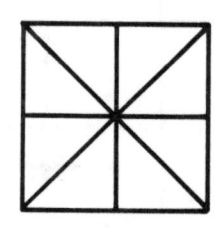

"seven of twelve" = $\frac{7}{}$

"one of two" = $\frac{}{2}$

"three of four" = $\frac{}{4}$

"two of five" = $\frac{}{5}$

"five of eight" = $\frac{5}{}$

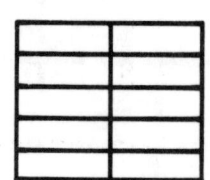

"nine of ten" = $\frac{9}{}$

Name _____   Skill: Writing fractions

There are 5 equal parts. → $\dfrac{3}{5}$ ← 3 parts are shaded.

Write the fraction for the following **shaded** parts.

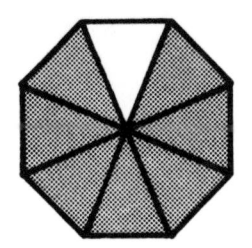

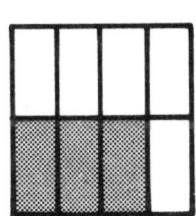

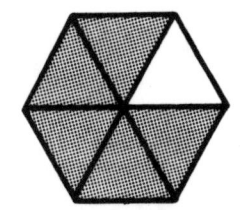

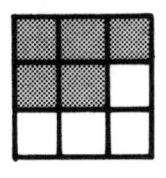

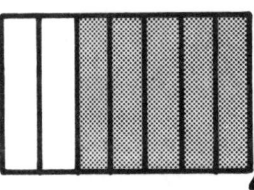

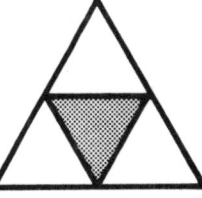

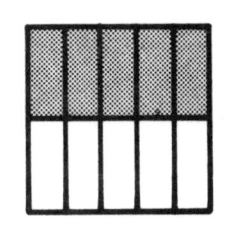

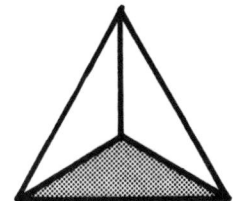

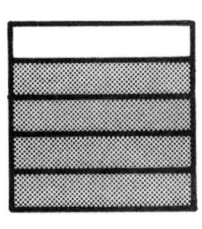

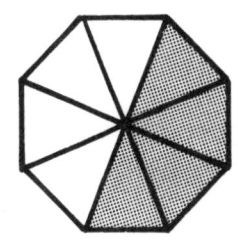

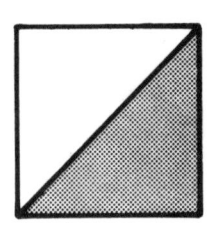

Use These Fractions

$\dfrac{1}{2}$ $\dfrac{1}{3}$ $\dfrac{2}{3}$ 　　　　　　　$\dfrac{2}{8}$ $\dfrac{3}{8}$ $\dfrac{4}{8}$

$\dfrac{1}{4}$ $\dfrac{2}{4}$ $\dfrac{3}{4}$ $\dfrac{3}{5}$ $\dfrac{4}{5}$ $\dfrac{1}{6}$ $\dfrac{4}{6}$ $\dfrac{5}{6}$ $\dfrac{5}{7}$ $\dfrac{7}{8}$ $\dfrac{5}{9}$ $\dfrac{5}{10}$

© Frank Schaffer Publications, Inc.　　　　75　　　　FS-32022 Math Activities

Name _____   Skill: Recognizing parts of a set

A fraction may be part of a group.

○ Two of the circles are shaded. → 2
◐ There are 3 circles in all. → ─
● Two-thirds are shaded.   3

Color as directed and answer the questions.

Color 3 of the 5 sea horses.

The fraction colored is _____ .

Color 5 of the 8 horseflies.

What fraction is colored? _____

Color 3 of the 4 ponytails.

What fraction is colored? _____

Color 4 of the 7 horseradish bottles.

What fraction is colored? _____

Color 5 of the 6 horseshoes.

What fraction is colored? _____

I start the day with 5 bags of hay.

© Frank Schaffer Publications, Inc.   FS-32022 Math Activities

Name _____   Skill: **Writing fractions in words**

1 of 3 puppies is dark.

One-third

Write these fraction names under the matching pictures: two-thirds, two-fifths, four-fifths, three-fourths, five-sixths, one-half.

Name _____   Skills: Parts of a whole, Parts of a set

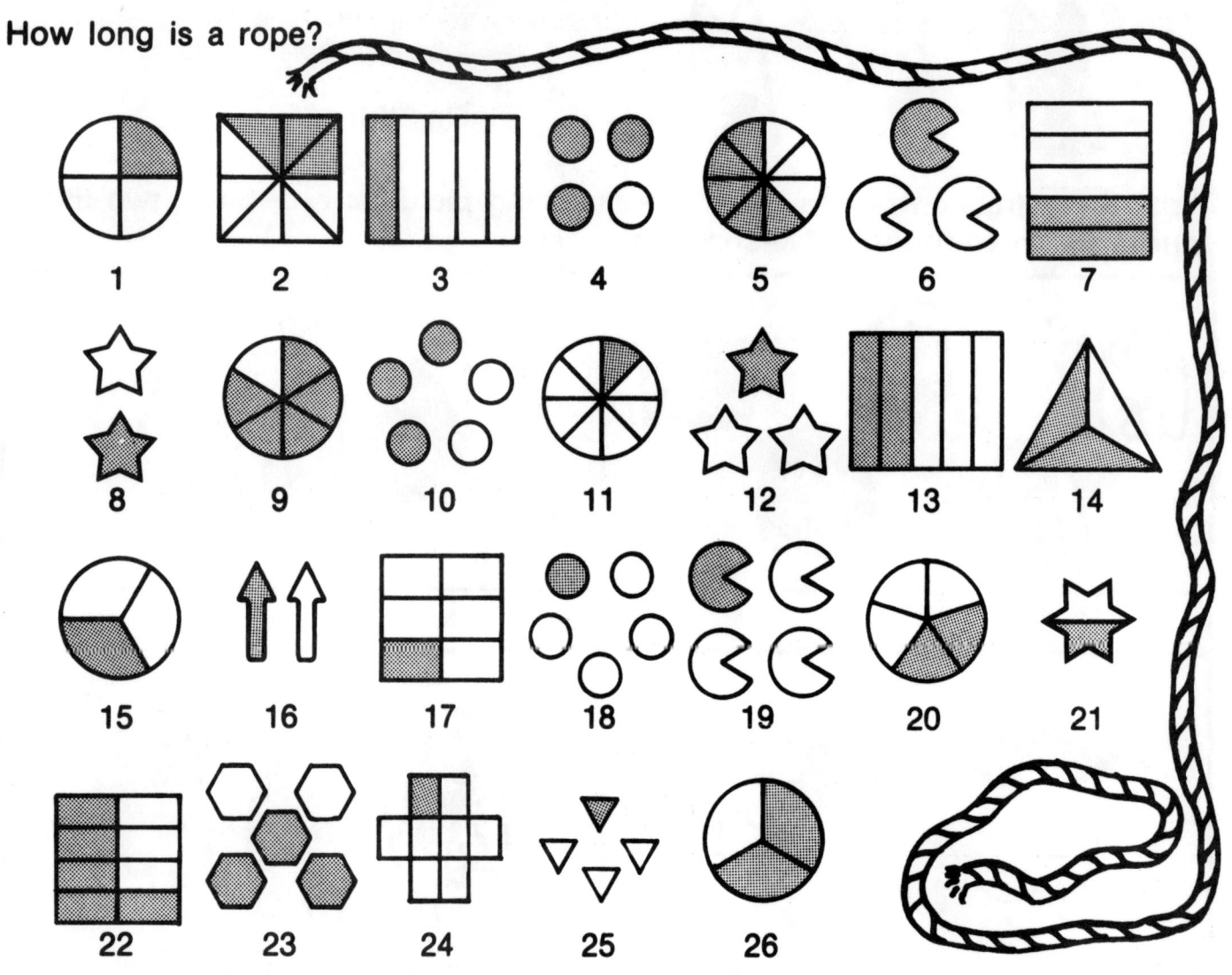

How long is a rope?

Find the letter that matches the fraction for each numbered figure above. Put the letters in the spaces below to solve the riddle. For example, if the shaded part of figure 1 is ¼, then **t** goes in the first space.

a	c	e	f	g	h	i	l	n	o	s	t	w
$\frac{1}{3}$	$\frac{3}{4}$	$\frac{5}{8}$	$\frac{1}{6}$	$\frac{1}{8}$	$\frac{2}{3}$	$\frac{1}{5}$	$\frac{1}{2}$	$\frac{3}{5}$	$\frac{5}{6}$	$\frac{2}{5}$	$\frac{1}{4}$	$\frac{3}{8}$

$\frac{t}{1}$ $\frac{}{2}$ $\frac{}{3}$ $\frac{}{4}$ $\frac{}{5}$ $\frac{}{6}$ $\frac{}{7}$ $\frac{}{8}$ $\frac{}{9}$ $\frac{}{10}$ $\frac{}{11}$ $\frac{}{12}$ $\frac{}{13}$

$\frac{}{14}$ $\frac{}{15}$ $\frac{}{16}$ $\frac{}{17}$ $\frac{}{18}$ $\frac{}{19}$ $\frac{}{20}$ $\frac{}{21}$ $\frac{}{22}$ $\frac{}{23}$ $\frac{}{24}$ $\frac{}{25}$ $\frac{}{26}$

© Frank Schaffer Publications, Inc.   FS-32022 Math Activities

Name _____   Skill: Number lines with 2-5 equal parts

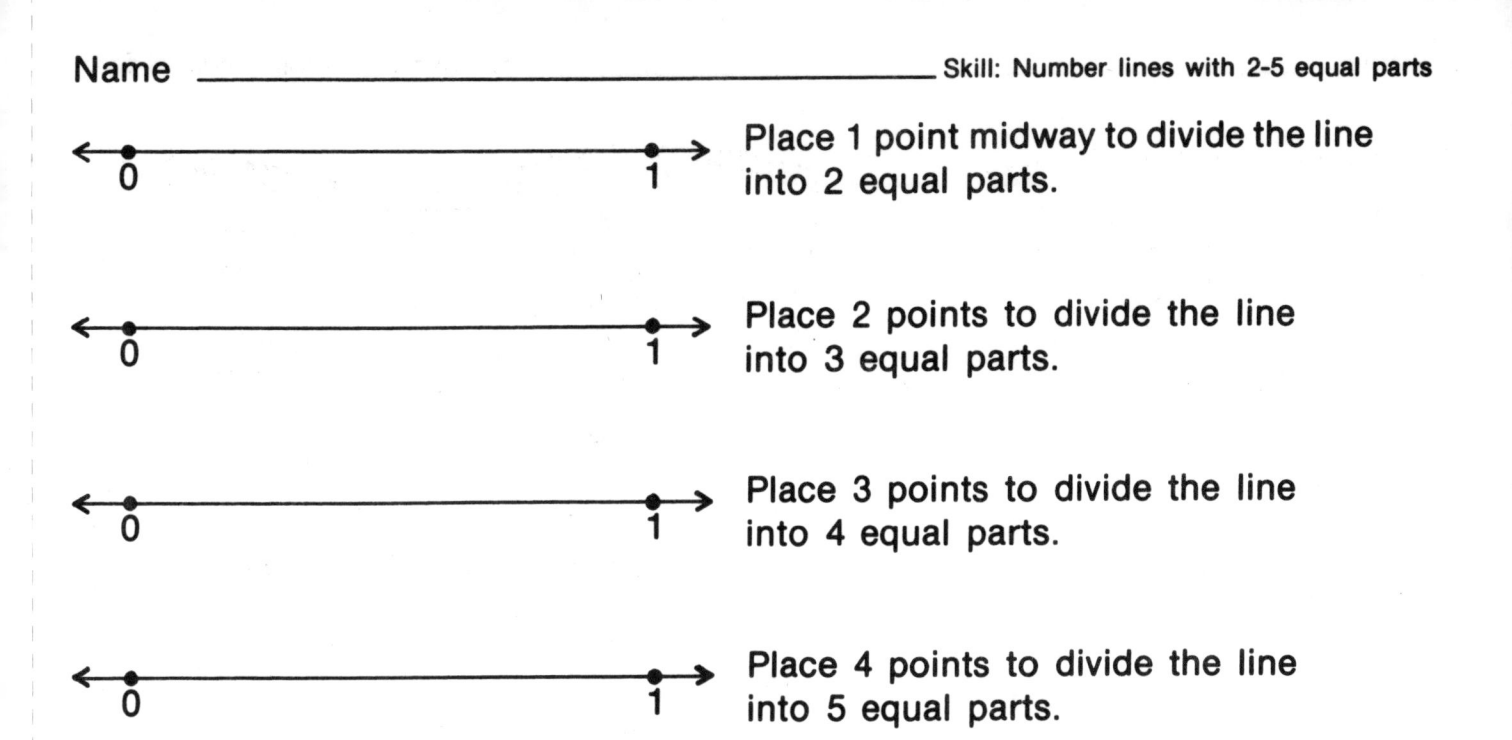

Place 1 point midway to divide the line into 2 equal parts.

Place 2 points to divide the line into 3 equal parts.

Place 3 points to divide the line into 4 equal parts.

Place 4 points to divide the line into 5 equal parts.

How far has each racer gone? Fill in the blanks.

Answers

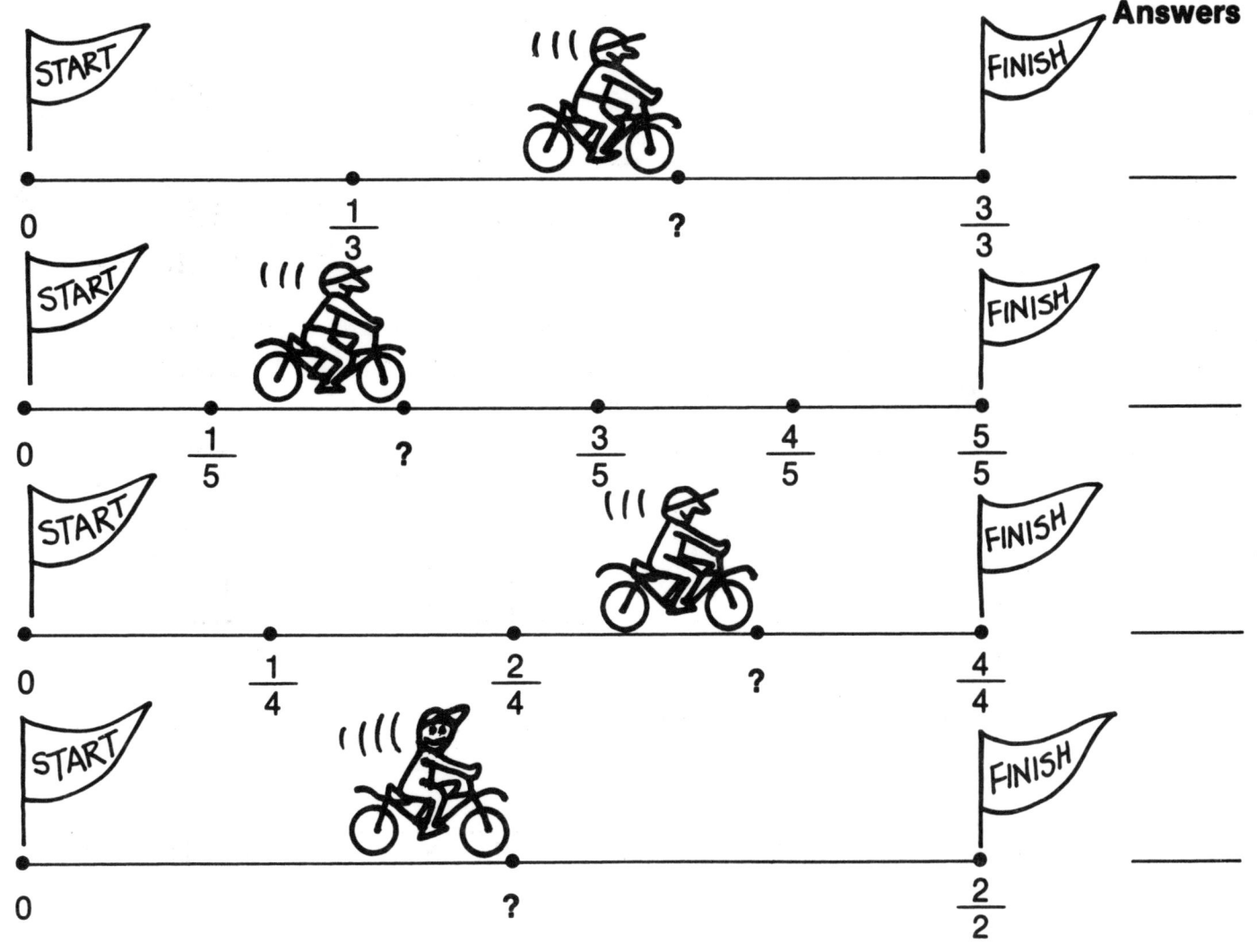

Racer 1: 0 — $\frac{1}{3}$ — ? — $\frac{3}{3}$ _____

Racer 2: 0 — $\frac{1}{5}$ — ? — $\frac{3}{5}$ — $\frac{4}{5}$ — $\frac{5}{5}$ _____

Racer 3: 0 — $\frac{1}{4}$ — $\frac{2}{4}$ — ? — $\frac{4}{4}$ _____

Racer 4: 0 — ? — $\frac{2}{2}$ _____

© Frank Schaffer Publications, Inc.   79   FS-32022 Math Activities

Name _____   Skill: Number lines with 6-8 equal parts

Place 5 points to divide the line into 6 equal parts.

Place 6 points to divide the line into 7 equal parts.

Place 7 points to divide the line into 8 equal parts.

How far has each fish swum to safety? Fill in the blanks.

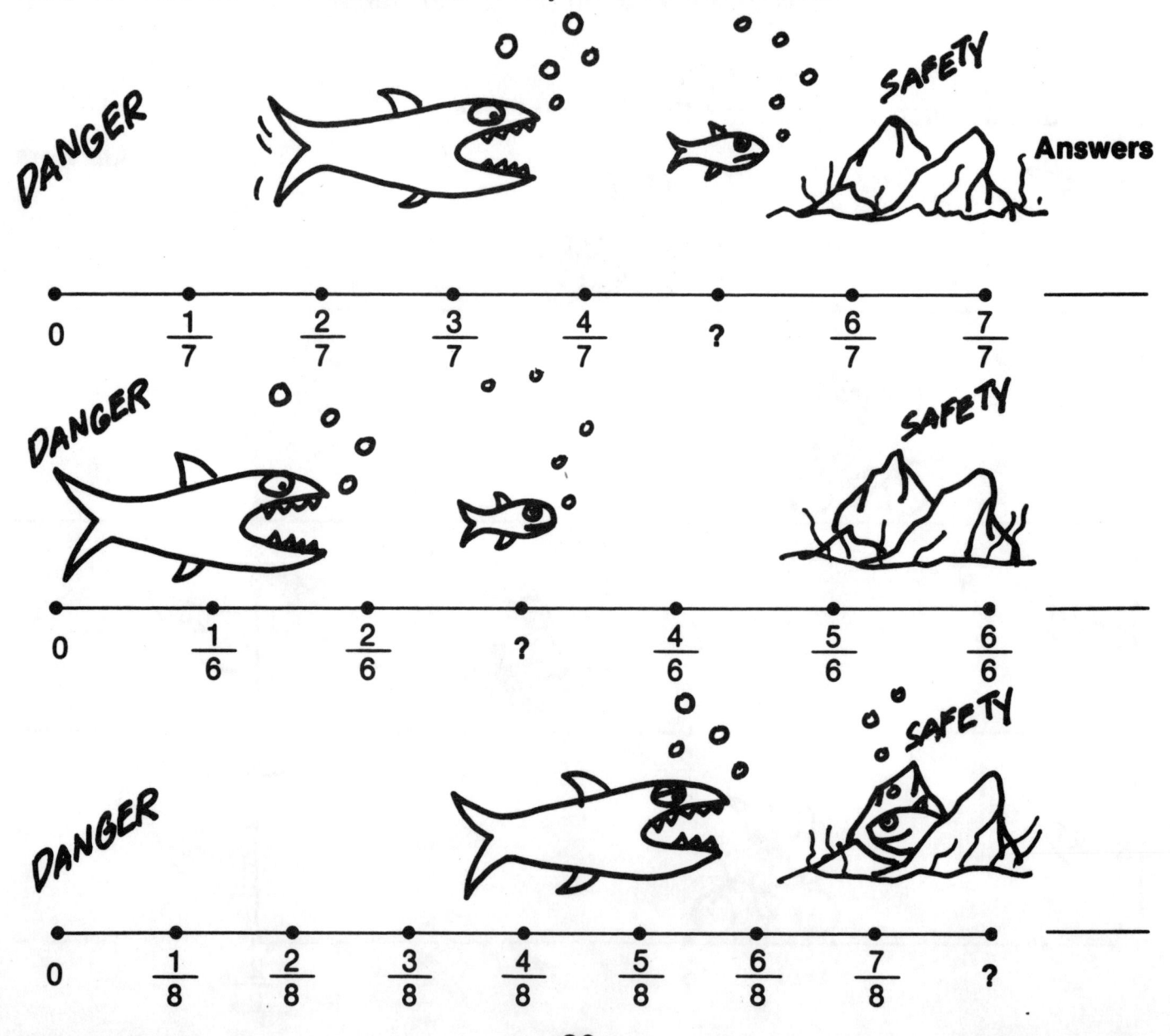

0   $\frac{1}{7}$ $\frac{2}{7}$ $\frac{3}{7}$ $\frac{4}{7}$? $\frac{6}{7}$ $\frac{7}{7}$

0   $\frac{1}{6}$ $\frac{2}{6}$? $\frac{4}{6}$ $\frac{5}{6}$ $\frac{6}{6}$

0   $\frac{1}{8}$ $\frac{2}{8}$ $\frac{3}{8}$ $\frac{4}{8}$ $\frac{5}{8}$ $\frac{6}{8}$ $\frac{7}{8}$?

Name _____ Skill: Number lines

←●————————————————●→ Place 3 points to divide the line
0   1   into 4 equal parts.

←●————————————————●→ Place 7 points to divide the line
0   1   into 8 equal parts.

←●————————————————●→ Place 2 points to divide the line
0   1   into 3 equal parts.

←●————————————————●→ Place 5 points to divide the line
0   1   into 6 equal parts.

How far has each climber gone? Fill in the blanks.

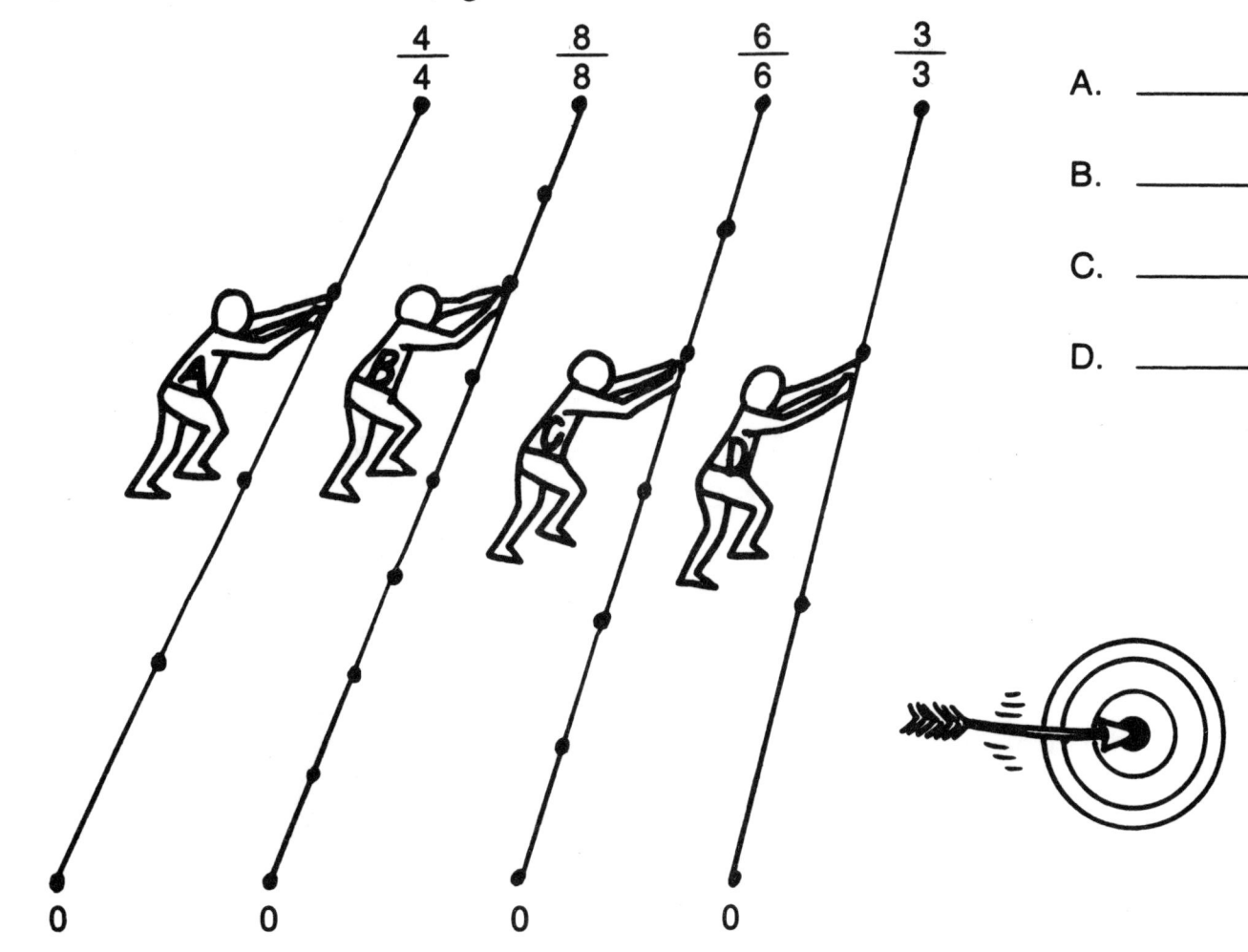

$\frac{4}{4}$ $\frac{8}{8}$ $\frac{6}{6}$ $\frac{3}{3}$

A. _____

B. _____

C. _____

D. _____

0   0   0   0

© Frank Schaffer Publications, Inc.   81   FS-32022 Math Activities

Name _____   Skill: Number lines

Find which gateway leads to fame!
Other paths are not the same.
Fill in the blanks and read the clues.
It's up to you to think and choose.

The fraction is:

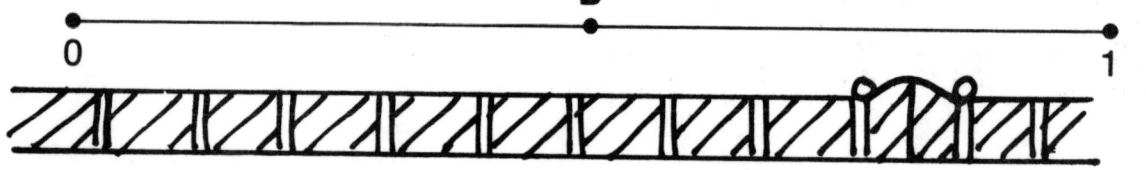

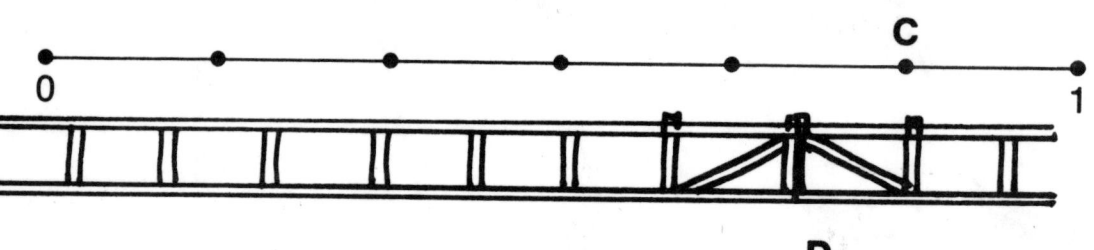

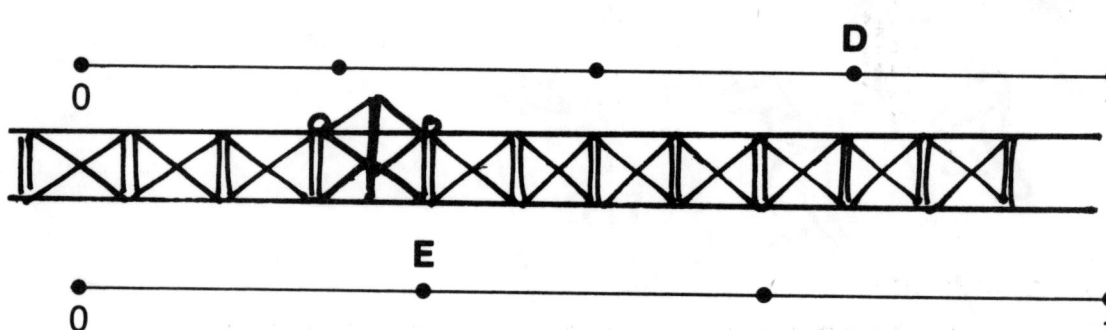

A _____

B _____

C _____

D _____

E _____

It's more than one-third, less than five-sixths,
More than two-thirds. There aren't any tricks!
What is your answer?
     GATEWAY _____

Name _____   Skill: Equal fractions

What part of each fence has holes?

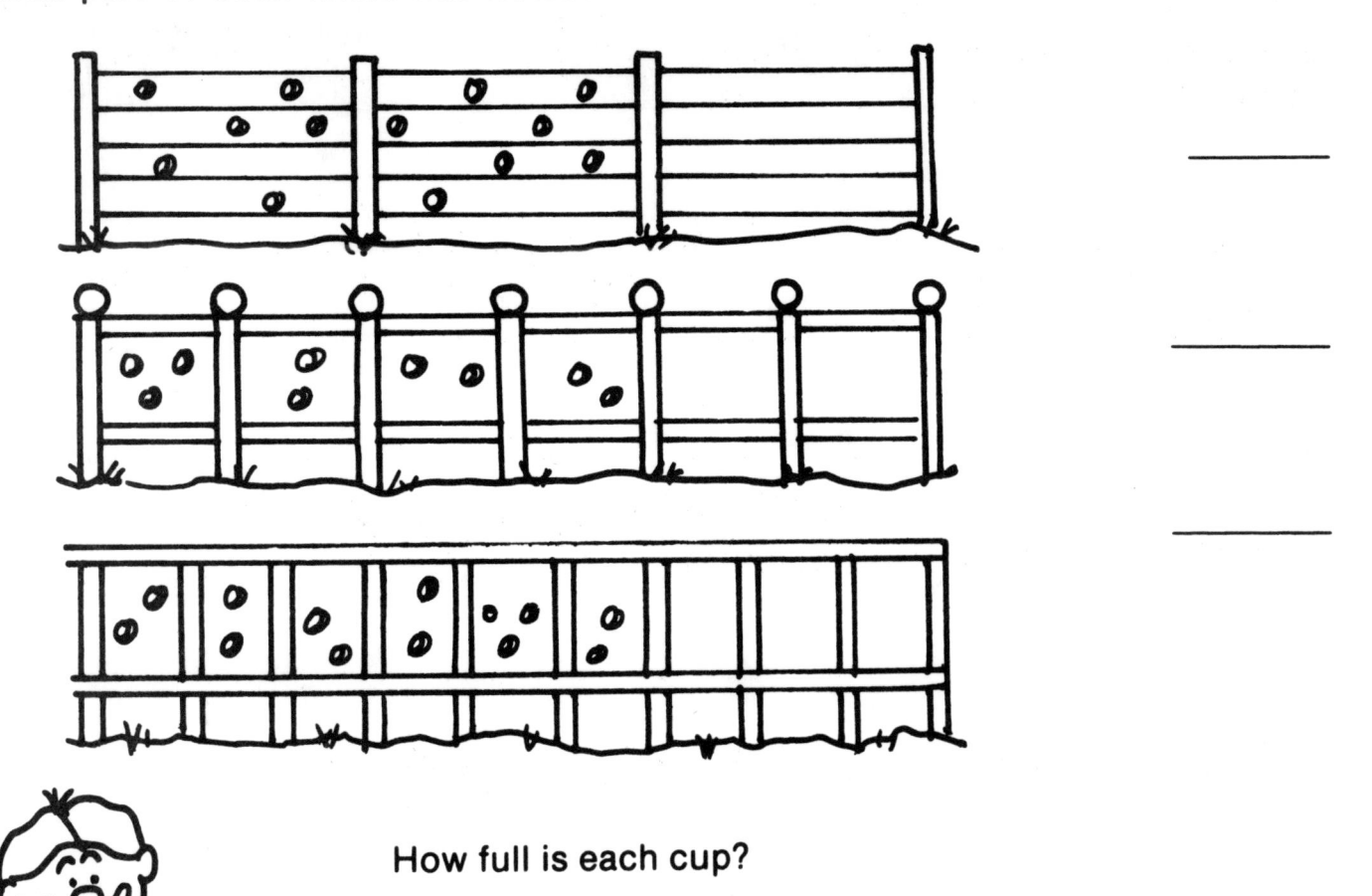

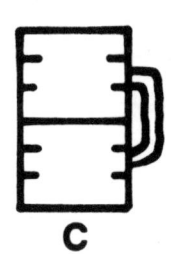

_____

_____

_____

How full is each cup?

A
B
C
D

A _____
B _____
C _____
D _____

What fraction is shaded?

A _____
B _____
C _____
D _____

83

Name _____   Skill: Equal fractions

How far has each bird flown?

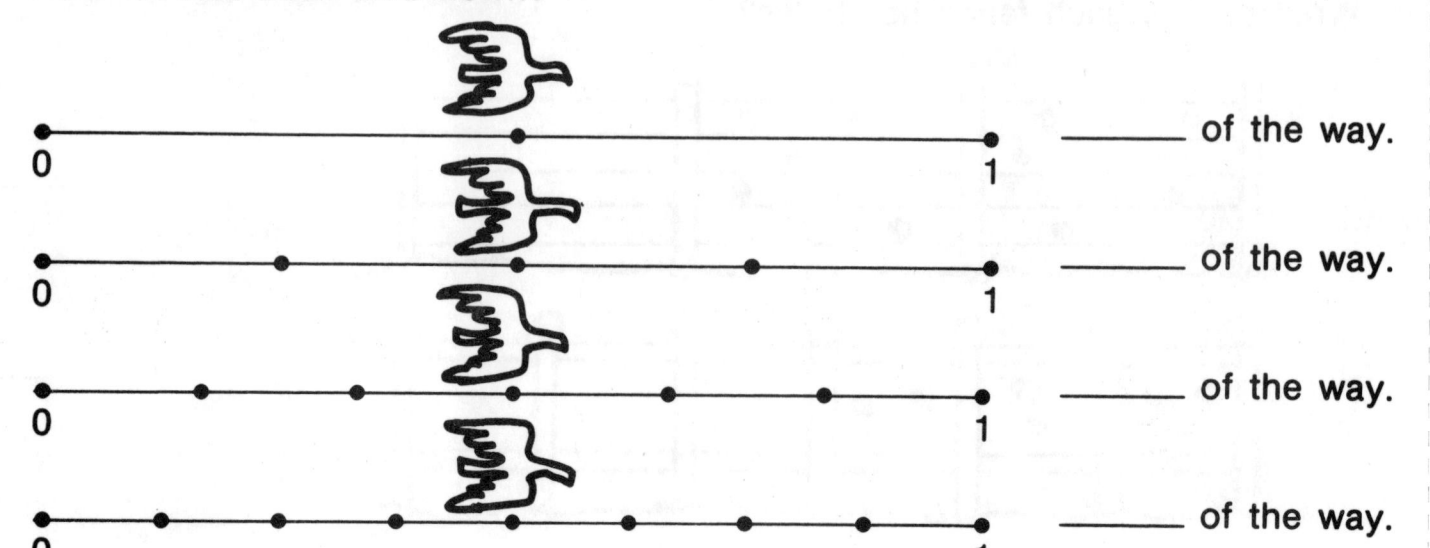

0 •———————•———————• 1   _____ of the way.

0 •———•———•———•———• 1   _____ of the way.

0 •———•———•———•———•———•———• 1   _____ of the way.

0 •—•—•—•—•—•—•—•—•—• 1   _____ of the way.

What portion of the penguins are lying down?

A   _____

B   _____

C   _____

What fraction is shaded?

A	$\frac{1}{3}$		$\frac{1}{3}$		$\frac{1}{3}$	
B	$\frac{1}{6}$	$\frac{1}{6}$	$\frac{1}{6}$	$\frac{1}{6}$	$\frac{1}{6}$	$\frac{1}{6}$
C	$\frac{1}{9}$ $\frac{1}{9}$ $\frac{1}{9}$		$\frac{1}{9}$ $\frac{1}{9}$ $\frac{1}{9}$		$\frac{1}{9}$ $\frac{1}{9}$ $\frac{1}{9}$	

A _____
B _____
C _____

Name _____ Skill: Equal fractions

What fraction of the balloons have popped?

_____   _____   _____

What fraction of the eggs have been dropped?

_____   _____   _____

What fraction is shaded?

A	1/4	1/4	1/4	1/4	A _____
B	1/8	1/8	1/8	1/8	1/8	1/8	1/8	1/8	B _____
C	1/12	1/12	1/12	1/12	1/12	1/12	1/12	1/12	1/12	1/12	1/12	1/12	C _____

© Frank Schaffer Publications, Inc.   85   FS-32022 Math Activities

Name _____   Skill: Equal fractions

Write the correct fraction words under each shaded figure.

two-eighths	three-ninths	one-third
two-fourths	one-fourth	two-sixths
three-twelfths	three-sixths	one-half

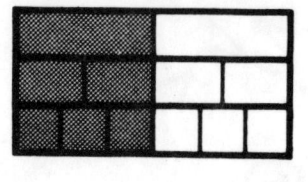

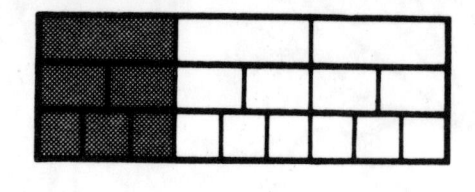

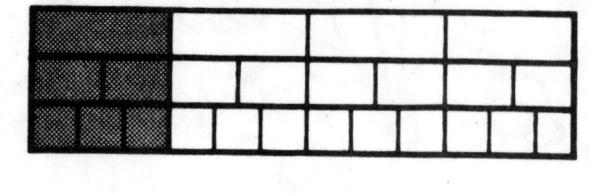

_____   _____   _____
_____   _____   _____
_____   _____   _____

Put the letter of each fraction or shaded figure under the fraction word below that names the same amount. Unscramble the letters to make a surprise sentence.

one-half	one-third	one-fourth

F $\dfrac{2}{4}$ R    N    R $\dfrac{1}{2}$ A

E 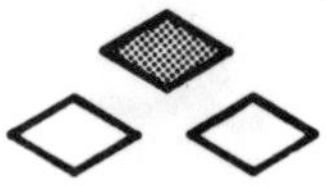   U $\dfrac{3}{12}$ C 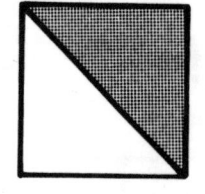   N 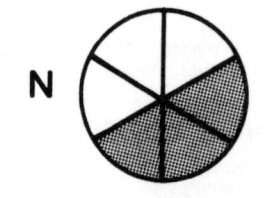   A

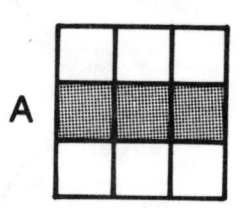

F 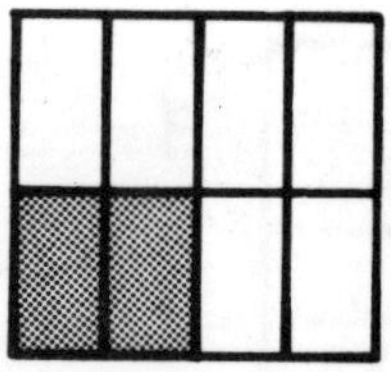   T    S 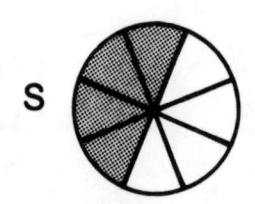   O $\dfrac{3}{6}$ I

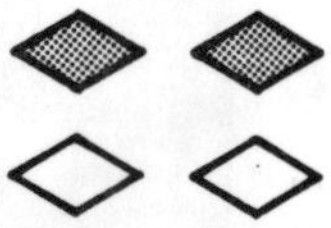

© Frank Schaffer Publications, Inc.   FS-32022 Math Activities

Name _____ Skill: Recognizing equal fractions

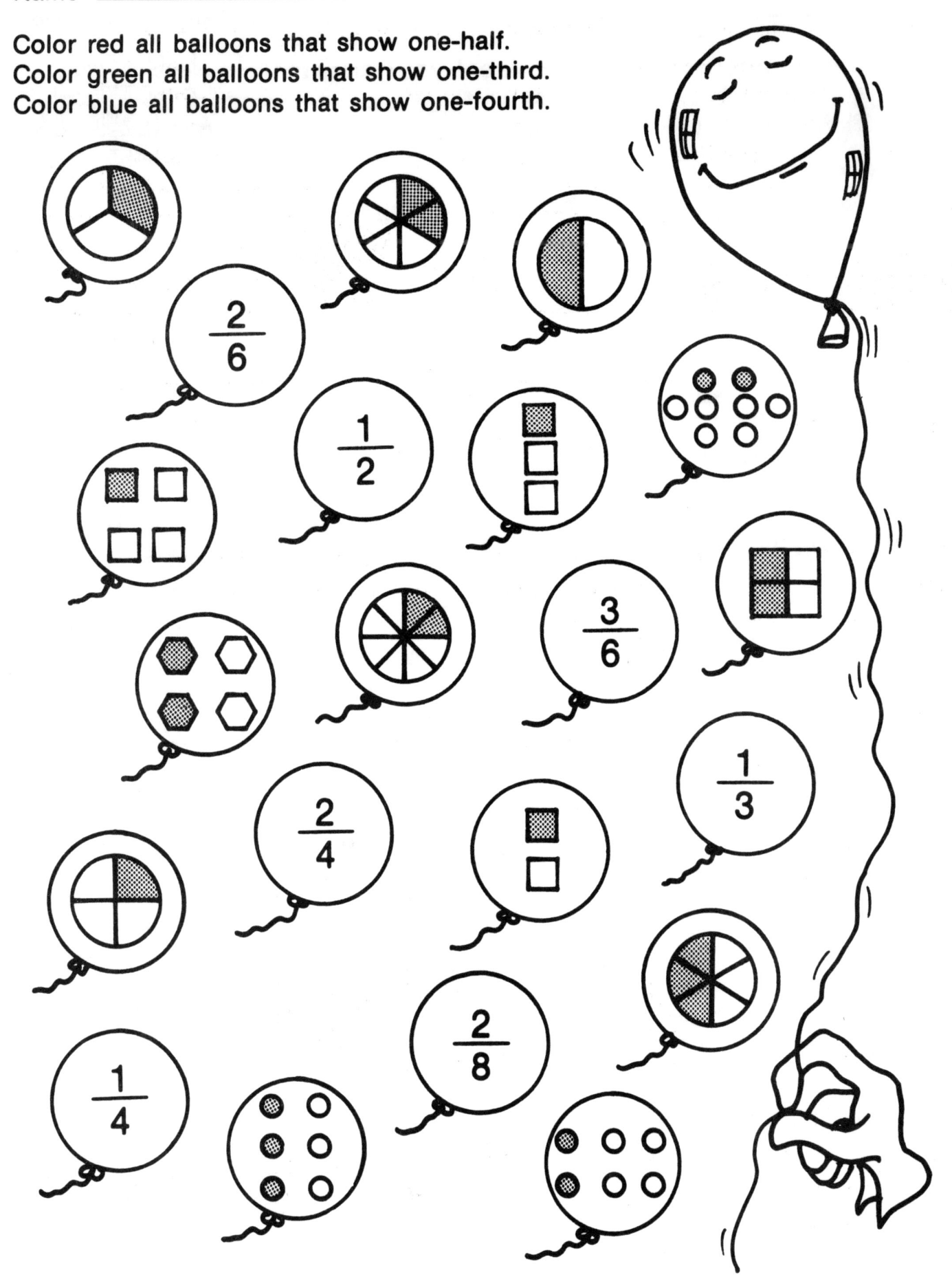

Color red all balloons that show one-half.
Color green all balloons that show one-third.
Color blue all balloons that show one-fourth.

Name _____   Skill: 3 digit addition

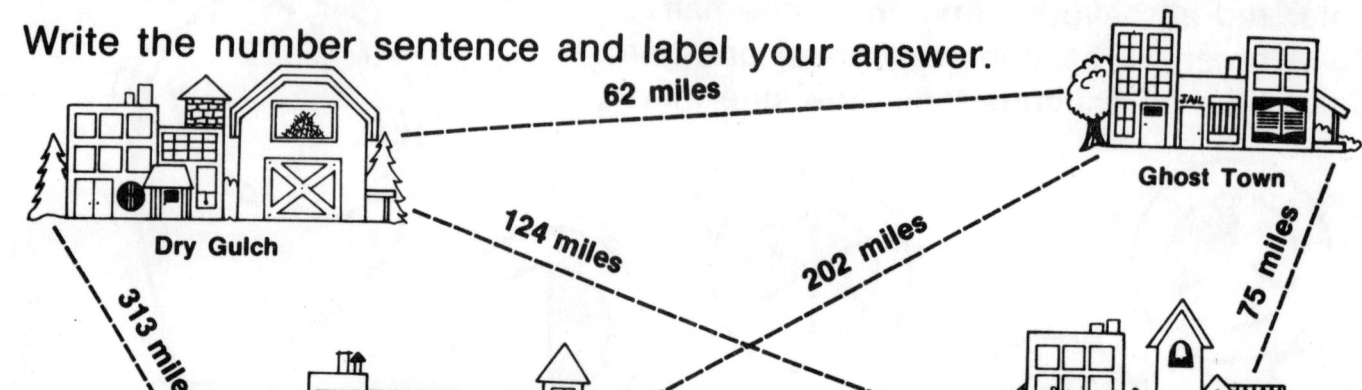

Write the number sentence and label your answer.

Dry Gulch — 62 miles — Ghost Town
Dry Gulch — 124 miles — Miners City
Ghost Town — 202 miles — Strike it Rich
Ghost Town — 75 miles — Miners City
Dry Gulch — 313 miles — Strike it Rich
Strike it Rich — 151 miles — Miners City

not to scale

1. The Browns drove from Dry Gulch to Strike it Rich. Then they drove to Miners City the next day. How many miles did they drive?

   313 + 151 = 464

   They drove **464 miles**.

2. From Miners City, the Browns headed for Ghost Town. Then they went back to Strike it Rich. How many miles did they travel?

   _____

   They traveled _____ _____.

3. The Williams family drove from Ghost Town to Dry Gulch, and then on to Miners City. How many miles did they travel?

   _____

   They traveled _____ _____.

4. From Miners City, the Williamses drove to Strike it Rich, and then to Ghost Town. How far did they go?

   _____

   They went _____ _____.

5. How many miles is it from Strike it Rich to Ghost Town, to Dry Gulch, and then to Miners City?

   _____

   It is _____ _____.

6. How long a trip would it be from Strike it Rich to Dry Gulch, to Miners City, and back to Strike it Rich?

   _____

   It would be _____ _____.

Name _____ Skills: Comparisons (3 digit)

Write the number sentence and label your answer.

1. Jan's airplane tickets cost her $463. Sue's were $586. How much more did Sue spend than Jan?

586−463=123

Sue spent ___$123___ more.

2. Bill flew 375 miles on Monday and 130 miles on Tuesday. How much farther did he fly on Monday than Tuesday?

_____

Bill flew _____ more.

3. Captain Barns flies at 678 mph. Captain Frost flies at 361 mph. How much faster is Captain Barns' plane?

_____

His plane is _____ faster.

4. Jack flew 264 miles. Mary flew 784. How many more miles did Mary fly?

_____

Mary flew _____ more.

5. A trip to Hawaii is $898. A skiing trip is $465. How much more is the trip to Hawaii?

_____

Hawaii is _____ more.

6. A 747 airplane can seat 366 people. A DC-10 can carry 260 people. How many more people can fly on a 747?

_____

_____ more.

© Frank Schaffer Publications, Inc.   89   FS-32022 Math Activities

Name _____ Skill: 3 digit addition and subtraction

Write the number sentence and label your answer.

RICE POP. 472
UPTON POP. 215
BLAIR POP. 300
NEWARK POP. 987

1. How many people live in the towns of Rice and Upton combined?

_____

There are _____ _____ .

2. If 250 people move out of Newark, how many will be left?

_____

_____ _____ will be left.

3. If all the people in Blair and Upton get together, how many would there be?

_____

There would be _____ _____ .

4. If 131 of the people in Rice go on vacation, how many will still be there?

_____

_____ _____ will still be there.

5. If 102 people in Upton move, how many will be left?

_____

_____ _____ will be left.

6. How many people live in Upton, Blair, and Rice combined?

_____

There are _____ _____ .

Name _____   Skill: 3 digit addition-regrouping

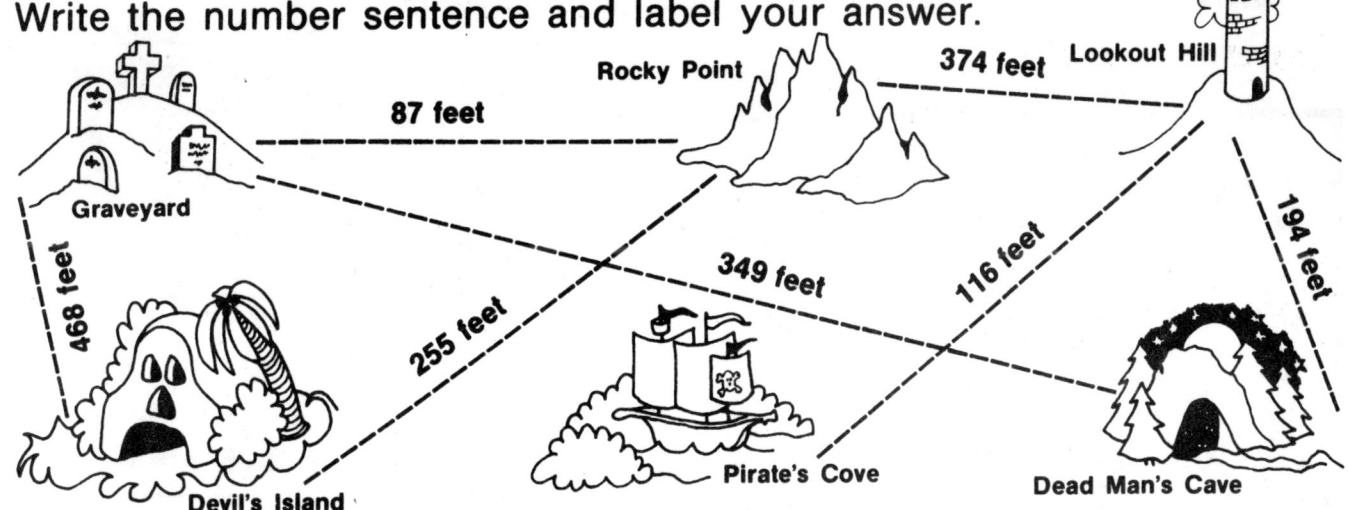

Write the number sentence and label your answer.

Rocky Point — 87 feet — Graveyard
Rocky Point — 374 feet — Lookout Hill
Graveyard — 468 feet — Devil's Island
Devil's Island — 255 feet — Pirate's Cove
Pirate's Cove — 349 feet — Rocky Point
Pirate's Cove — 116 feet — Dead Man's Cave
Dead Man's Cave — 194 feet — Lookout Hill

not to scale

1. Sue and Bill started at Dead Man's Cave, walked to the Graveyard and then to Devil's Island. How far did they go?

   349 + 468 = 817

   They walked __817 feet__.

2. John and Jim went from Dead Man's Cave to Lookout Hill and then to Pirate's Cove. How far had they gone?

   _____

   They went _____.

3. If Nancy and Ann walked from Devil's Island to Rocky Point and then to the Graveyard, how long was their trip?

   _____

   Their trip was _____ long.

4. Tom started at Devil's Island and headed for the Graveyard. From there, he walked to Rocky Point. How many feet did he walk?

   _____

   Tom walked _____.

5. How far is it from Lookout Hill to Rocky Point and back again?

   _____

   It is _____.

6. How many feet is it from Rocky Point to the Graveyard if you go by Devil's Island?

   _____

   It is _____.

Name _____ Skill: 3 digit-regrouping—comparisons

Write the number sentence and label your answer.

hippo	924 pounds
bear	826 pounds
gorilla	342 pounds
zebra	325 pounds
lion	267 pounds
seal	129 pounds
wolf	102 pounds

1. How much more does the hippo weigh than the lion?

 924 − 267 = 657

 __657__ __pounds__ more.

2. How much less does the wolf weigh than the gorilla?

 _____

 _____ _____ less.

3. How much difference in weight is there between the zebra and the seal?

 _____

 The difference is ____ _____ .

4. How much difference in weight is there between the bear and the gorilla?

 _____

 The difference is ____ _____ .

5. How much less does the bear weigh than the hippo?

 _____

 _____ _____ less.

6. How much more does the gorilla weigh than the lion?

 _____

 _____ _____ more.

Name _____   Skills: Addition and Subtraction—2-3 Digits with Regrouping

Sports of all Sorts!

Write an equation and label your answer.

1. Stan Superstar did 210 situps in 3 minutes. Allie Ace did 185. How many more situps did Stan S. do than Allie A.?

   _____

2. In last night's basketball game, Marty made 23 points, Mary made 19, Millie and Missie each made 15. Susie made 3. How many points did they make all together?

   _____

3. 609 soccer fans came to see the big game. 352 sat on one side. How many soccer fans sat on the other side of the field?

   _____

4. Baron Bounder hit the baseball 132 feet into center field. If the center fielder was standing 95 feet from home plate, how far did he have to run to get the ball?

   _____

5. Sara Sidesaddle rode her pinto pony, Peanuts, 245 yards on Trail A, then 145 yards on Trail B. How far did Sara and Peanuts go?

   _____

*6. Gary Gutter and Sam Splitz bowled two games at Trendy Ten Pins. S.S. scored 97 and 132. G.G. scored 102 and 115. How many more pins did S.S. knock down?

   _____

   _____

Name _____

Skills: Addition and Subtraction—3-4 Digits with Regrouping

How Inventive are Inventors?

Write an equation and label your answer.

1. Professor Twitcher invented 168 wacky ways to wake a person up and 303 wacky ways to put them to sleep. How many wacky ways did he invent all together?

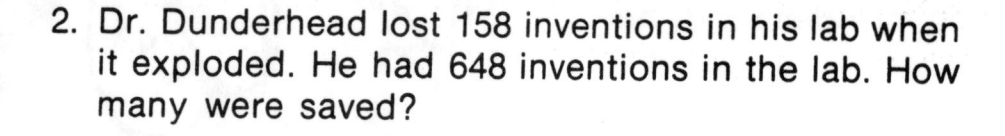

   _____

2. Dr. Dunderhead lost 158 inventions in his lab when it exploded. He had 648 inventions in the lab. How many were saved?

   _____

3. The safety pin was invented in 1849. The zipper was invented 42 years later. In what year was the zipper invented?

   _____

4. Edison patented more than 1100 inventions in his lifetime. He lived from 1847-1931. How old was he when he died?

   _____

5. 400 inventors entered a contest to make jet-powered canoes. 218 canoes stayed afloat. How many sank?

   _____

6. Super Smart Sally sold her design for a singing swing for $2,500. Her sister Sue sold her design for silent skates for $1,800. How much did they earn all together?

   _____

© Frank Schaffer Publications, Inc.

FS-32022 Math Activities

Name _____ Skill: Multiplication

Write the number sentence and label your answer.

1. There are 3 spiders. If each spider has 8 legs, how many legs are there in all?

   3 × 8 = 24

   There are __24__ __legs__.

2. There are 3 dogs. If each dog has 4 legs, how many legs are there altogether?

   _____

   There are _____ _____.

3. If there are 5 ants, each with 6 legs, how many legs are there?

   _____

   There are _____ _____.

4. Ducks have 2 feet. If there are 7 ducks, how many feet are there in all?

   _____

   There are _____ _____.

5. An octopus has 8 arms. How many arms would there be if you had 4 octopuses?

   _____

   There would be _____ _____.

6. Snakes don't have legs. If there are 3 snakes, how many legs are there?

   _____

   There are _____ _____.

7. If one cricket has 6 legs, how many legs would 3 crickets have altogether?

   _____

   There would be _____ _____.

Name _____ Skill: Multiplication

Write the number sentence and label your answer.

1. A football team gets 6 points for a touchdown. How many points would 8 touchdowns be?

_____

It would be _____ _____ .

2. Baseball games have 9 innings. If each inning has 3 outs, how many outs are there in a game?

_____

There are _____ _____ .

3. If you swam 8 laps a day for 7 days, how many laps would you swim altogether?

_____

That would be _____ _____ .

4. There are 4 people at each ping pong table. If there are 9 tables, how many people are playing?

_____

There are _____ _____ .

5. Dad jogs 6 miles a day. How many miles would he jog in 4 days?

_____

Dad would jog _____ _____ .

6. There are 5 players on a basketball team. If they each score 9 points, what would the team's score be?

_____

The score would be _____ .

7. In tennis, you play 6 games each set. If you play 7 sets, how many games would you play?

_____

You would play _____ _____ .

Name _____ Skill: Division

Write the number sentence and label your answer.

1. There were 16 boy scouts. They slept in 8 tents. How many boys were in each tent?

__ _____

Each tent had _____ _____ .

2. If 32 girl scouts stayed in 4 cabins, how many were in each cabin?

_____

Each cabin had _____ _____ .

3. There are 6 large tables in the mess hall. If there are 54 scouts, how many will sit at each table?

_____

_____ _____ at each table.

4. 63 girl scouts wanted to go boating. If there are 9 boats, how many girls will be in each boat?

_____

_____ _____ in each boat.

5. Seven groups of boy scouts went hiking. There were 42 boys. How many were in each group?

_____

Each group had _____ _____ .

6. If 25 scouts make 5 campfires, how many will sit at each campfire?

_____

_____ _____ at each campfire.

7. The arts and crafts leader can take groups of 9 scouts at a time. If 27 scouts sign up, how many groups would come?

_____

There would be _____ _____ .

Name _____   Skill: Multiplication or division

Write the number sentence and label your answer.

1. Jack planted 3 rows of apple trees, with 6 trees in each row. How many apple trees does he have?

_____

Jack has ____ _____ .

2. Mary divided 10 peaches into 2 baskets. How many peaches are in each basket?

_____

____ _____ in each basket.

3. If Sam planted 21 orange trees in 7 rows, how many trees are in each row?

_____

Each row had ____ _____ .

4. Sue had 9 baskets. If she put 7 apples in each one, how many apples did she pick?

_____

Sue picked ____ _____ .

5. Bill had 7 peach trees. He picked 4 good peaches from each tree. How many peaches did he have in all?

_____

Bill had ____ _____ .

6. John divided 30 oranges equally between 6 friends. How many oranges did each friend get?

_____

Each friend got ____ _____ .

7. Betty has 63 apples to put into 7 baskets. How many apples did she put in each basket?

_____

Each basket had ____ _____ .

Name _____ Skill: Multiplication or division

Write the number sentence and label your answer.

1. Jim has 36¢. If candy bars are 9¢ each, how many can he buy?

_____

Jim can buy ____ _____ .

2. Sue has 7¢. Gumdrops are 1¢ each. How many can she buy?

_____

Sue can buy ____ _____ .

3. Bill bought 5 jawbreakers for 4¢ each. How much did he spend?

_____

Bill spent _____ .

4. Candy mints are 3¢ each. If Nancy has 6¢, how many can she buy?

_____

Nancy can buy ____ _____ .

5. Jack bought 2 packages of gum. There are 8 pieces of gum in each package. How many pieces did he have?

_____

Jack had ____ _____ .

6. Bob wants to buy 6 candy bars. They are 9¢ each. How much money does he need?

_____

Bill needs _____ .

7. The candy man has 8 jars of candy sticks. Each jar has 5 sticks of candy. How many are there in all?

_____

There are ____ _____ .

Name _____

How Amusing are Amusement Parks?

Write an equation and label your answer.

1. Ted, Todd, Ann and Amanda went to Gulp and Gasp Park last Saturday. They had tickets for 36 rides and divided them equally. How many tickets did each child get?

   _____

2. They each bought a G & G poster priced at 15¢ apiece. How much did they spend for the 4 posters?

   _____

3. They met 3 boys who had each ridden The Spine-Tingler 14 times. How many times is that all together?

   _____

4. Each car of the Whirling Dervish held 3 people. 28 people wanted to ride. Only one didn't get on. How many cars were there?

   _____

5. The kids had a total of $16 to spend on lunch and other things. If they divided it by 4, how much would each child get to spend?

   _____

6. By lunchtime, Ted, Todd and Ann had each gone on 5 rides. Amanda had gone on 4 rides. How many tickets had they used?

   _____

7. How many tickets did they have left?

   _____

Name _____

I Love a Parade!

Write an equation and label your answer.

1. The Turkey Bowl Parade will have 8 bands. If there are 10 in each band, how many band members will be marching?

   _____

2. There will be 9 rows to sit in along the way. Each row can hold 9 people. How many people can sit in all nine rows?

   _____

3. 110 people are coming to see the parade. How many will not be able to sit in the rows?

   _____

4. The drill team from Tempo High has 29 girls. One is a baton twirler. The rest will march in 4 rows. How many girls will be in each row?

   _____

5. How many rows of horses can be lined up 8 across if there are 56 horses?

   _____

6. If there are 9 floats each carrying 8 turkeys and 7 floats each carrying 9 ducks, which floats are carrying more birds: the turkey float or the duck float? How many more?

   _____

   _____

© Frank Schaffer Publications, Inc.   FS-32022 Math Activities

Name _____

From Little Wheels to Big Wheels

Write an equation and label your answer.

1. If one pair of roller skates costs $36, how much do 4 pairs cost?

   _____

2. Dizzy Turner did 5 "360's" in a row on his skateboard. If there are 360 degrees in a circle, how many degrees did he turn all together?

   _____

3. It took Tony Tenspeed 2 days to travel 68 miles on his bike. If he went the same distance each day, how many miles did he pedal per day?

   _____

4. If a go-cart could go 27 miles per hour, how far could it go in 5 hours?

   _____

5. Robert rode in his Rabbit from Redfield to Greenfield, a distance of 99 miles. It took 3 hours. How fast was he driving?

   _____

6. Abe drove his eighteen-wheeler for 4 hours at 54 mph. Sid drove his sixteen-wheeler for 5 hours at 45 mph. How many more miles did Sid drive than Abe?

   _____

   _____

Name _____

Glub for Gold

Write an equation and label your answer.

1. 2 deep-sea divers spent $350 each to search for sunken treasure. How much did they spend all together?

   _____

2. Freddy Fin found the hulk of a galleon in 6 hours after diving and searching for 306 feet. How many feet per hour is that?

   _____

3. 8 treasure chests were found, each containing 90 coins. How many coins were found?

   _____

4. If 7 divers, 1 tender, and 1 captain shared these coins equally, how many coins did each keep?

   _____

5. Each coin was worth $8. If they sold all the coins, how much would each one get?

   _____

6. How much money did each diver earn on this trip?

   _____

Name _____

Skills: Additon and Subtraction of Like Fractions

Bunk on a Trunk

Write an equation and label your answer.

1. The tree house was $\frac{1}{5}$ mile east of Tom's house and $\frac{2}{5}$ mile west of Tony's house. How far was it from Tom's to Tony's?

　_____

2. Tom cleaned the tree house $\frac{1}{6}$ of an hour on Sunday and $\frac{4}{6}$ of an hour on Monday. How long did Tom clean?

　_____

3. Terry tried to sneak up the rope ladder. She got $\frac{9}{10}$ of the way and slipped $\frac{6}{10}$ down. Where did that leave her?

　_____

4. Barbie brought them $\frac{3}{4}$ of a pizza. If she ate $\frac{1}{4}$, what did that leave Tom and Tony?

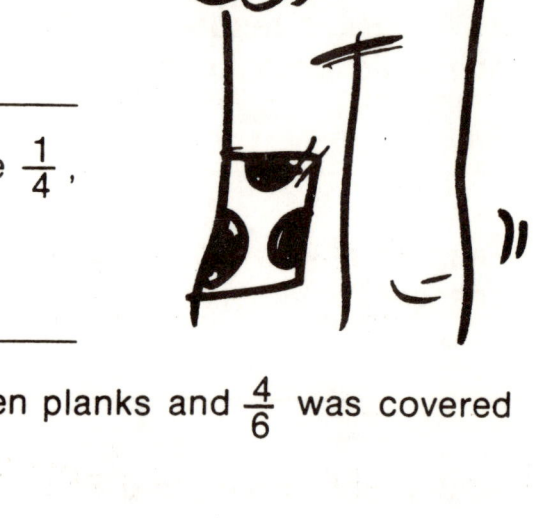

　_____

5. $\frac{1}{6}$ of the tree house was covered with wooden planks and $\frac{4}{6}$ was covered with a sheet. How much was covered?

　_____

6. Their gallon bucket of water had $\frac{1}{8}$ gal. of water left in it. They lowered it by rope and Barbie poured $\frac{5}{8}$ of a gal. of water into it. On the way up, $\frac{3}{8}$ gal. spilled. How much was left?

© Frank Schaffer Publications, Inc.　　　　104　　　　FS-32022 Math Activities

Name _____   Skills: Addition and Subtraction of Like Fractions

And Four to Go!

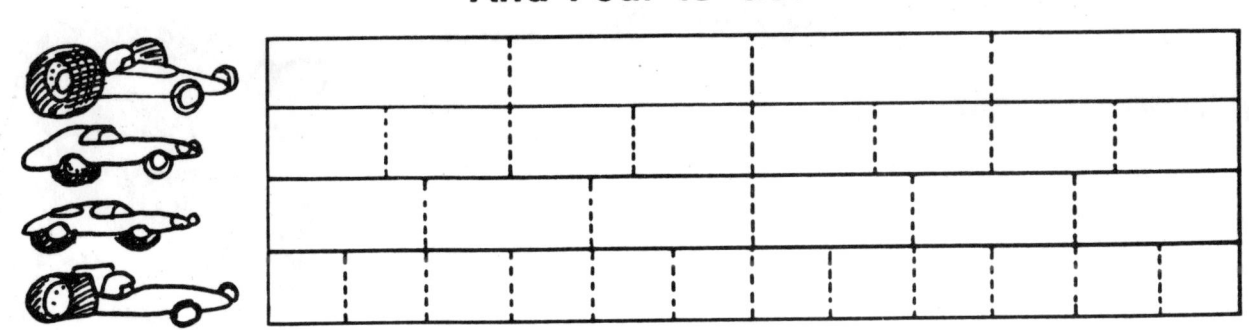

Draw the cars on the racetrack according to your answers.

Write an equation and label your answer.

1. Andy's slot car drove $\frac{1}{4}$ of the way on his first turn and $\frac{1}{4}$ of the way on his second turn. How far had it gone?

   _____

2. Bob's slot car drove $\frac{3}{8}$ of the way on his first turn and $\frac{3}{8}$ on his second turn. How far had it gone?

   _____

3. How much farther had Bob's car gone than Andy's?

   _____

4. Cindy's slot car drove $\frac{1}{6}$ of the way on her first turn and $\frac{1}{6}$ of the way on her second turn. How far had it gone?

   _____

5. Doris' slot car drove $\frac{5}{12}$ of the way on her first turn and $\frac{4}{12}$ of the way on her second turn. How far had it gone?

   _____

6. Which two cars were in the lead? _____

   Which car was the farthest behind? _____

© Frank Schaffer Publications, Inc.   FS-32022 Math Activities

Name _____

Skills: Addition and Subtraction of Like Fractions

Obnoxious Concoctions

Write an equation and label your answer.

1. Cora added $\frac{1}{4}$ cup ginger ale to $\frac{2}{4}$ cup castor oil. How much does she have?

   _____

2. Add $\frac{2}{5}$ cup flour to $\frac{1}{5}$ cup salt and $\frac{1}{5}$ cup water. How much goop will you have?

   _____

3. Give $\frac{3}{5}$ cup of your goop to a goop-lover. What's left?

   _____

4. Brenda blended $\frac{2}{8}$ cup butter to $\frac{3}{8}$ cup bitter batter and $\frac{2}{8}$ cup buttermilk. How many cups did she blend?

   _____

5. Karl's cure for colds is $\frac{3}{8}$ cup hot water, $\frac{1}{8}$ cup lemon juice, $\frac{3}{8}$ cup Tabasco sauce. How much does this make?

   _____

6. Jerry mixed $\frac{7}{8}$ cup olive juice with $\frac{7}{8}$ cup pickle juice. He drank $\frac{7}{8}$ of a cup and threw out the rest. How much did he throw out?

   _____

© Frank Schaffer Publications, Inc.   FS-32022 Math Activities

Name _____ Skill: Review Test

12	18	20	24	30
green	red	orange	brown	yellow

107

© Frank Schaffer Publications, Inc. FS-32022 Math Activities

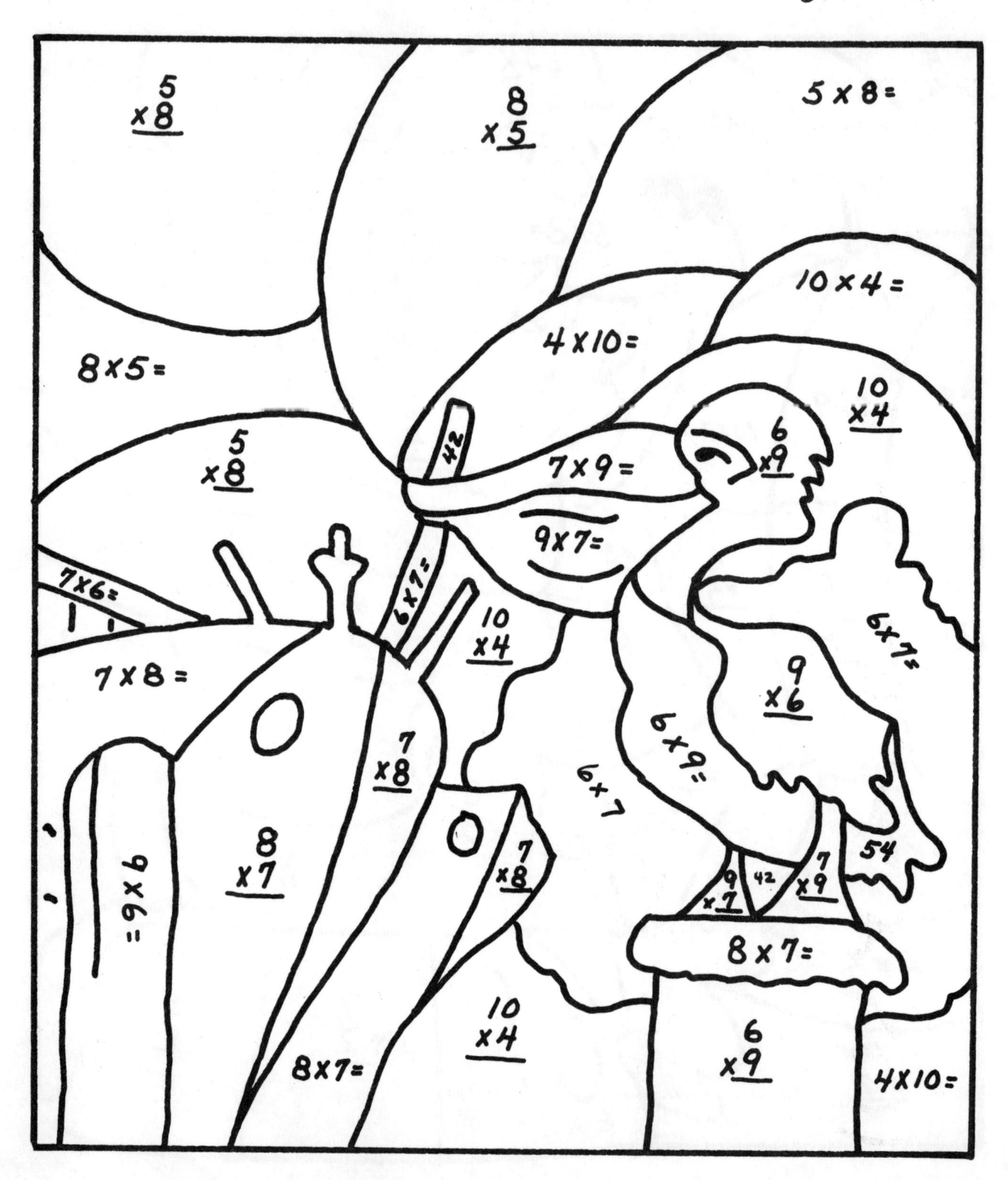

Name _____

Skill: Review Test

40	42	54	56	63
blue	orange	brown	red	yellow

108

Name _____

Skill: review

6	7	8	9	10
yellow	green	blue	red	orange

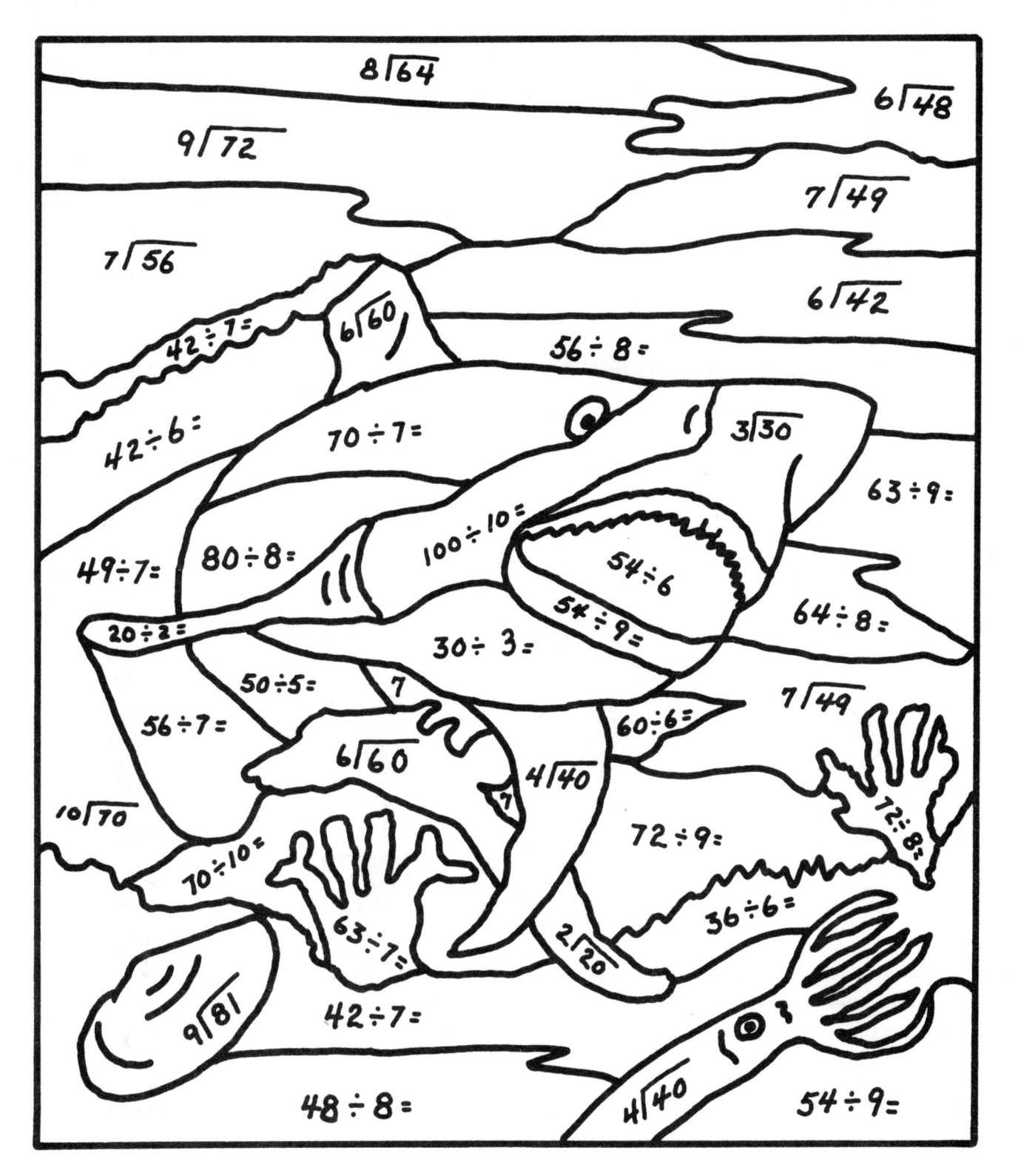

8)64
6)48
9)72
7)49
7)56
6)42
42÷7=
6)60
56÷8=
42÷6=
70÷7=
3)30
63÷9=
49÷7=
80÷8=
100÷10=
54÷6
64÷8=
20÷2=
30÷3=
54÷9=
50÷5=
7
7)49
56÷7=
60÷6=
6)60
4)40
72÷9=
10)70
72÷8=
70÷10=
36÷6=
63÷7=
2)20
9)81
42÷7=
48÷8=
4)40
54÷9=

109

Name _____   Pre/Post Test

I. How many equal parts does each figure have?

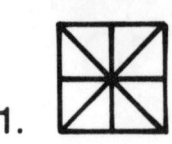

1. _____   2. _____   3. _____

Write the fraction for each word.

4. one-half _____   5. two-thirds _____

II. Put the letter of each fraction next to the right figure.

A. five-sixths   D. $\frac{2}{5}$
B. one-third
C. three-fourths   E. $\frac{3}{8}$

III. Write a fraction for each letter.

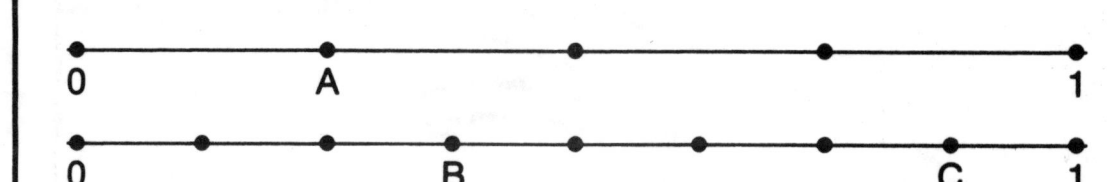

A. _____
B. _____
C. _____

IV. Put a check (✔) in front of all words, figures and fractions that are the same as one-half.

_____ $\frac{3}{6}$ _____ $\frac{2}{4}$ _____ $\frac{1}{3}$

_____ two-fourths
_____ two-fifths
_____ five-tenths

SCORE	I	II	III	IV	TOTAL ___ X 4 = ___ %
Possible Right:	(5)	(5)	(3)	(12)	(Possible Right: 25)

Answer Key

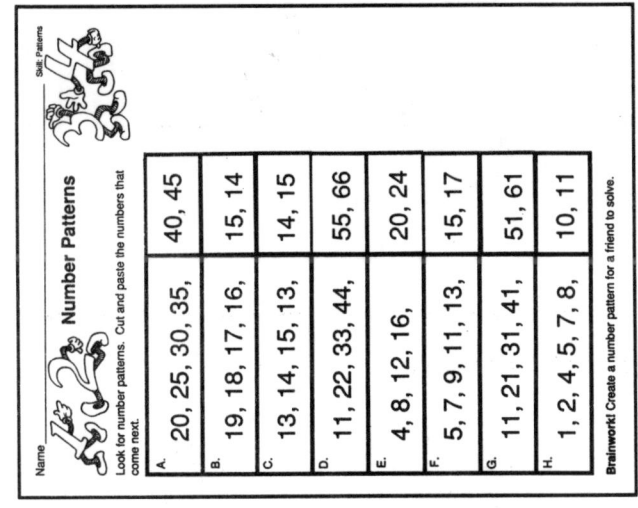

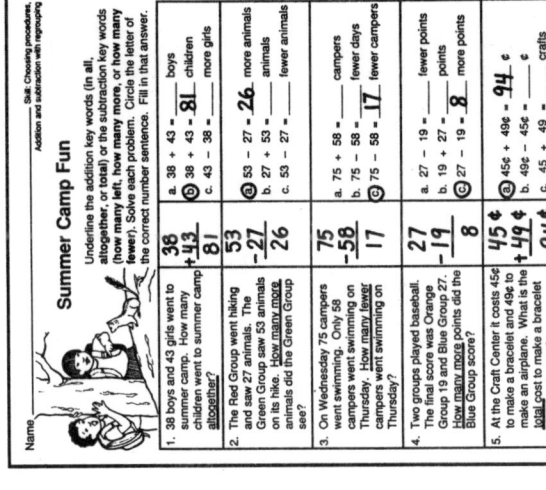

Page 3 — Number Patterns

Look for number patterns. Cut and paste the numbers that come next.

A.	20, 25, 30, 35,	40, 45
B.	19, 18, 17, 16,	15, 14
C.	13, 14, 15, 13,	14, 15
D.	11, 22, 33, 44,	55, 66
E.	4, 8, 12, 16,	20, 24
F.	5, 7, 9, 11, 13,	15, 17
G.	11, 21, 31, 41,	51, 61
H.	1, 2, 4, 5, 7, 8,	10, 11

Brainwork! Create a number pattern for a friend to solve.

Page 6 — Summer Camp Fun

Underline the addition key words (in all, altogether, or total) or the subtraction key words (how many left, how many more, or how many fewer). Solve each problem. Circle the letter of the correct number sentence. Fill in that answer.

1. 38 boys and 43 girls went to summer camp. How many children went to summer camp altogether?
   a. 38 + 43 = __ boys
   ⓑ 38 + 43 = 81 children
   c. 43 − 38 = __ more girls
   38 + 43 = 81

2. The Red Group went hiking and saw 27 animals. The Green Group saw 53 animals on its hike. How many more animals did the Green Group see?
   ⓐ 53 − 27 = 26 more animals
   b. 27 + 53 = __
   c. 53 − 27 = __ fewer animals
   53 − 27 = 26

3. On Wednesday 75 campers went swimming. Only 58 campers went swimming on Thursday. How many fewer campers went swimming on Thursday?
   a. 75 + 58 = __ campers
   b. 75 − 58 = __ fewer days
   ⓒ 75 − 58 = 17 fewer campers
   75 − 58 = 17

4. Two groups played baseball. The final score was Orange Group 19 and Blue Group 27. How many more points did the Blue Group score?
   a. 27 − 19 = __ fewer points
   b. 19 + 27 = __ points
   ⓒ 27 − 19 = 8 more points
   27 − 19 = 8

5. At the Craft Center it costs 45¢ to make a bracelet and 49¢ to make an airplane. What is the total cost to make a bracelet and an airplane?
   ⓐ 45¢ + 49¢ = 94¢
   b. 49¢ − 45¢ = __ ¢
   c. 45 + 49 = __ crafts
   45¢ + 49¢ = 94¢

Brainwork! Would you add or subtract to find out how many more campers brought sleeping bags than sheets and blankets? Write a word problem about this. Ask a friend to solve it.

Page 2 — Shape Patterns

Look for patterns. Cut and paste the shapes that come next.

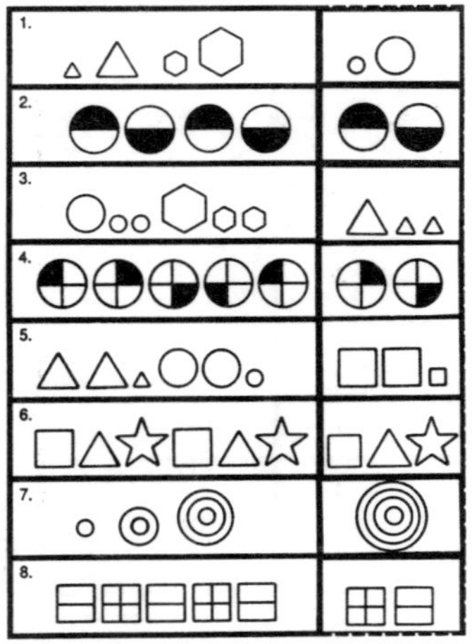

Brainwork! Draw a pattern using shapes. Ask a friend to look for the pattern and draw the shape that comes next.

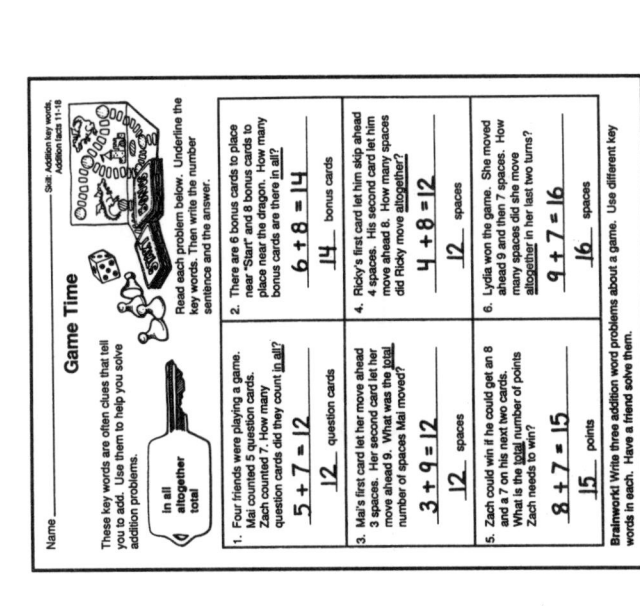

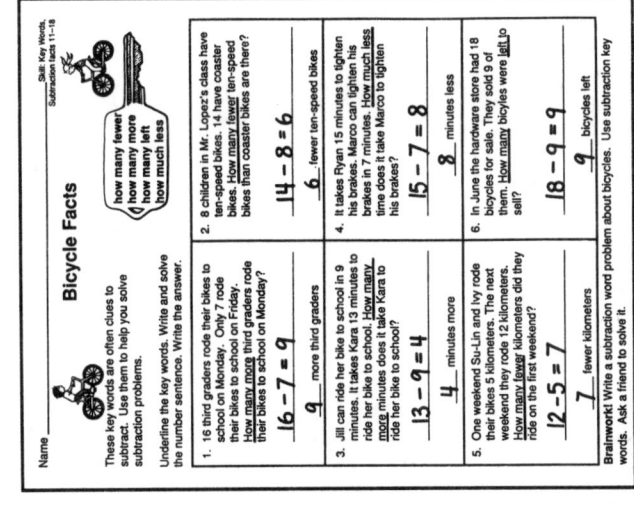

Page 5 — Bicycle Facts

These key words are often clues to subtract. Use them to help you solve subtraction problems.

how many fewer
how many more
how much less

Underline the key words. Write and solve the number sentence. Write the answer.

1. 16 third graders rode their bikes to school on Monday. Only 7 rode their bikes to school on Friday. How many more third graders rode their bikes to school on Monday?
   16 − 7 = 9
   9 more third graders

2. 8 children in Mr. Lopez's class have ten-speed bikes. 14 have coaster bikes. How many fewer ten-speed bikes than coaster bikes are there?
   14 − 8 = 6
   6 fewer ten-speed bikes

3. Jill can ride her bike to school in 9 minutes. It takes Kara 13 minutes to ride her bike to school. How many more minutes does it take Kara to ride her bike?
   13 − 9 = 4
   4 minutes more

4. It takes Ryan 15 minutes to tighten his brakes. Marco can tighten his brakes in 7 minutes. How much less time does it take Marco to tighten his brakes?
   15 − 7 = 8
   8 minutes less

5. One weekend Su-Lin and Ivy rode their bikes 5 kilometers. The next weekend they rode 12 kilometers. How many fewer kilometers did they ride on the first weekend?
   12 − 5 = 7
   7 fewer kilometers

6. In June the hardware store had 18 bicycles for sale. They sold 9 of them. How many bicycles were left to sell?
   18 − 9 = 9
   9 bicycles left

Brainwork! Write a subtraction word problem about bicycles. Use subtraction key words. Ask a friend to solve it.

Page 1 — Dancing Digits

Use the numbers 1, 3, and 9 to write a three-digit number on each line.
1. What is the largest number that can be made? **931**
2. What is the smallest number that can be made? **139**
3. What is the largest number that can be made with a 3 in the hundreds place? **391**
4. What is the smallest number that can be made with a 3 in the ones place? **193**
5. What number is greater than 193 but less than 391? **319**

Use the numbers 2, 4, 7, and 8 to write a four-digit number on each line.
6. What is the largest number that can be made? **8742**
7. What is the smallest number that can be made? **2478**
8. What is the largest number that can be made with a 4 in the thousands place? **4872**
9. What is the smallest number that can be made with a 7 in the hundreds place? **2748**
10. What number is greater than 7284 but less than 7482? **7428**

Brainwork! There are 24 different four-digit number combinations that can be made using the numbers 1, 5, 6, and 9. Write the six combinations that begin with a 9.

Page 4 — Game Time

These key words are often clues that tell you to add. Use them to help you solve addition problems.

in all
altogether
total

Read each problem below. Underline the key words. Write the number sentence and the answer.

1. Four friends were playing a game. Mai counted 5 question cards. Zach counted 7. How many question cards did they count in all?
   5 + 7 = 12
   12 question cards

2. There are 6 bonus cards to place near "Start" and 8 bonus cards to place near the dragon. How many bonus cards are there in all?
   6 + 8 = 14
   14 bonus cards

3. Mai's first card let her move ahead 3 spaces. Her second card let her move ahead 9. What was the total number of spaces Mai moved?
   3 + 9 = 12
   12 spaces

4. Ricky's first card let him skip ahead 4 spaces. His second card let him move ahead 8. How many spaces did Ricky move altogether?
   4 + 8 = 12
   12 spaces

5. Zach could win if he could get an 8 and a 7 on his next two cards. What is the total number of points Zach needs to win?
   8 + 7 = 15
   15 points

6. Lydia won the game. She moved ahead 9 and then 7 spaces. How many spaces did she move altogether in her last two turns?
   9 + 7 = 16
   16 spaces

Brainwork! Write three addition word problems about a game. Use different key words in each. Have a friend solve them.

© Frank Schaffer Publications, Inc. 111 FS-32022 Math Activities

Answer Key

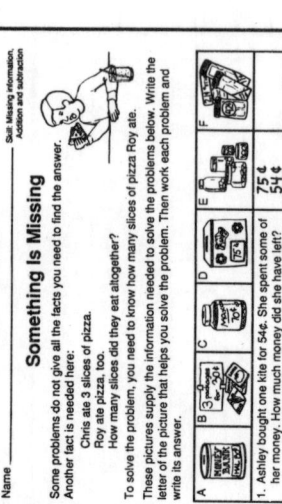

Page 9 — Something Is Missing

Some problems do not give all the facts you need to find the answer. Another fact is needed here:
Chris ate 3 slices of pizza.
Roy ate pizza, too.
How many slices did they eat altogether?

To solve the problem, you need to know how many slices of pizza Roy ate. These pictures supply the information needed to solve the problems below. Write the letter of the picture that helps you solve the problem. Then work each problem and write its answer.

1. Ashley bought one kite for 54¢. She spent some of her money. How much money did she have left? Which picture has the missing information? **D**
   75¢ − 54¢ = 21¢

2. Lynn used all of her money to buy 4 packages of stickers. How much money did she spend? Which picture has the missing information? **F**
   21¢ × 4 = 84¢ → **85¢**

3. Rita has saved 50¢. Hal has saved money, too. How much more money has Hal saved than Rita? Which picture has the missing information? **A**
   80¢ − 50¢ = 30¢ more

4. Carla saves pennies. She has 5 pennies in each jar. How many pennies does she have in all? Which picture has the missing information? **E**
   5 + 5 + 5 + 5 + 5 = **25 pennies**

5. On Friday Paco earned 75¢. He bought three seed packages. How much money does he have left? Which picture has the missing information? **B**
   75¢ − 30¢ = **45¢**

Brainwork! Max wants to buy baseball cards with his money. Each package costs 33¢. How many packages can he buy? Find the missing information above. Then solve the problem.

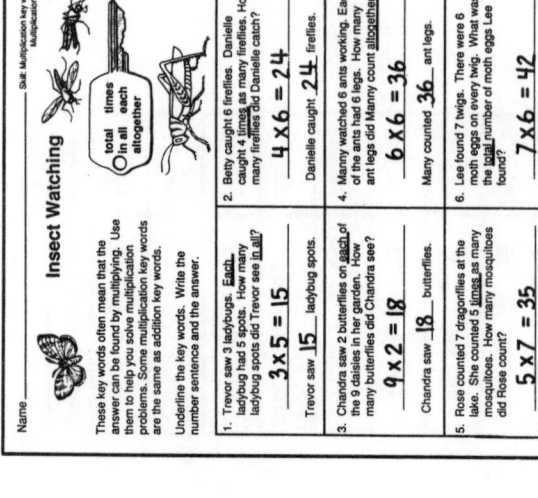

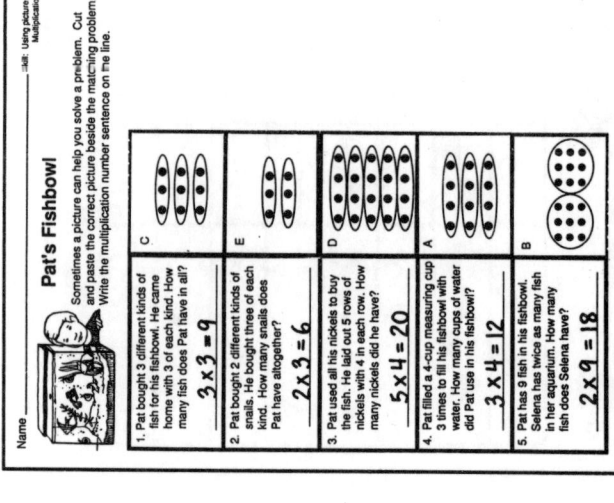

Page 12 — Insect Watching

These key words often mean that the answer can be found by multiplying. Use them to help you solve multiplication problems. Some multiplication key words are the same as addition key words.
Underline the key words. Write the number sentence and the answer.

total / times / in all / each / altogether

1. Trevor saw 3 ladybugs. Each ladybug had 5 spots. How many ladybug spots did Trevor see in all?
   3 × 5 = 15 ladybug spots

2. Betty caught 6 fireflies. Danielle caught 4 times as many fireflies. How many fireflies did Danielle catch?
   4 × 6 = 24 fireflies

3. Chandra saw 2 butterflies on each of the 9 daisies in her garden. How many butterflies did Chandra see?
   9 × 2 = 18 butterflies

4. Manny watched 6 ants working. Each of the ants had 6 legs. How many ant legs did Manny count altogether?
   6 × 6 = 36 ant legs

5. Rose counted 7 dragonflies at the lake. She counted 5 times as many mosquitoes. How many mosquitoes did Rose count?
   5 × 7 = 35 mosquitoes

6. Lee found 7 twigs. There were 6 moth eggs on every twig. What was the total number of moth eggs Lee found?
   7 × 6 = 42 moth eggs

Brainwork! Write a multiplication word problem about grasshoppers. Use multiplication key words. Have a friend solve it.

Page 8 — The Snack Shop

The Snack Shop needs your help. The cash registers are not working. Fill in the prices and total up the orders. Be sure to give back the correct change. Order A has been done for you.

Menu
- Superburger $.89
- Cheeseburger $.58
- Pizza Slice $.49
- Fruit Cup $.28
- Green Salad $.39
- Milk $.25
- Apple Juice $.55
- Yogurt $.35

A. Snack Shop
1 Yogurt $.35
1 Milk $.25
1 Fruit Cup $.28
Total Cost $.88
Money received $.95
Total Cost −$.88
Change $.07

B. Snack Shop
3 Yogurts $1.05
Total Cost $1.05
Money received $1.25
Total Cost −$1.05
Change $.20

D. Snack Shop
1 Superburger $.89
1 Milk $.25
1 Fruit Cup $.28
Total Cost $1.42
Money received $1.50
Total Cost −$1.42
Change $.08

C. Snack Shop
2 Superburgers $1.78
1 Fruit Cup $.28
1 Green Salad $.39
2 Apple Juice $1.10
Total Cost $3.55
Money received $3.75
Total Cost −$3.55
Change $.20

E. Snack Shop
1 Pizza Slice $.49
1 Milk $.25
1 Fruit Cup $.28
Total Cost $1.02
Money received $2.00
Total Cost −$1.02
Change $.98

Brainwork! Order lunch for yourself. Total the amount.

Page 11 — Pat's Fishbowl

Sometimes a picture can help you solve a problem. Cut and paste the correct picture beside the matching problem. Write the multiplication number sentence on the line.

1. Pat bought 3 different kinds of fish for his fishbowl. He came home with 3 of each kind. How many fish does Pat have in all?
   3 × 3 = 9 — C

2. Pat bought 2 different kinds of snails. He bought three of each kind. How many snails does Pat have altogether?
   2 × 3 = 6 — E

3. Pat used all his nickels to buy fish. He laid out 5 rows of nickels with 4 in each row. How many nickels did he have?
   5 × 4 = 20 — D

4. Pat filled a 4-cup measuring cup 3 times to fill his fishbowl with water. How many cups of water did Pat use in his fishbowl?
   3 × 4 = 12 — A

5. Pat has 9 fish in his fishbowl. Selena has twice as many fish in her aquarium. How many fish does Selena have?
   2 × 9 = 18 — B

Page 7 — Baseball Days

When solving a problem, first estimate what the answer will be.
1. Underline the key words.
2. Circle the most reasonable answer.
3. Write the number sentence for the exact answer.
4. Write the answer.
5. Check that the answer makes sense.

1. There are 21 boys and 30 girls in the league this summer. How many baseball players in all are in the league?
   About 10 (About 50) About 80
   19 + 30 = 51
   51 players

2. Mario's team practiced throwing for 8 minutes and batting for 19 minutes. How many more minutes did they spend practicing batting?
   (About 10) About 40
   19 − 8 = 11
   11 minutes

3. On Wednesday Mario's team scored 6 runs. Alicia's team scored 19 runs. How many fewer runs did Mario's team score than Alicia's?
   19 − 6 = 13 (About 10) About 50
   13 runs

4. Alicia's team practiced for 38 minutes on Friday and 60 minutes on Saturday. How many minutes did they practice altogether?
   About 50 (About 100)
   38 + 60 = 98
   98 minutes

5. Alicia hit the ball 57 feet on her first try and 78 feet on her next try. How many fewer feet did she hit the ball her first try than on her second try?
   (About 20) About 80
   78 − 57 = 21
   21 feet

6. Mario's team spent 23 dollars on shirts and 19 dollars on baseball hats. What was the total amount of money spent on shirts and hats?
   About 40 (About 70)
   23 + 19 = 42
   42 dollars

Brainwork! There are four bases on a baseball field. If Marco made 3 home runs, how many times did he touch a base? Estimate the answer, then solve the problem.

Page 10 — Shopping for School Supplies

Sometimes problems have too much information. Cross out any extra information. Add or subtract to solve each problem. Then write the answer.

1. Dad took Jessica and Troy shopping ~~for 3 hours~~. Jessica brought 7 quarters she had saved. Troy brought 8 quarters. How many quarters did Jessica and Troy bring in all?
   7 + 8 = 15 quarters

2. Jessica bought a sweatshirt on sale for 3 dollars. The matching slacks she bought cost 9 dollars. ~~Dad said not to spend more than 15 dollars.~~ How much did she spend altogether on her new outfit?
   3 + 9 = 12 dollars

3. It took Troy 14 minutes to buy a T-shirt and 6 minutes to buy socks. ~~He looked at books for 5 minutes.~~ How many more minutes did it take Troy to buy his new T-shirt than to buy his socks?
   14 − 6 = 8 minutes

4. Jessica bought ~~5 new notebooks~~, 16 pencils, and 7 erasers. How many more pencils than erasers did Jessica buy?
   16 − 7 = 9 more pencils

5. Jessica bought 2 boxes of markers. Each box had 9 markers. ~~Troy only wanted a box of crayons.~~ How many markers did she buy in all?
   2 × 9 = 18 or 9 + 9 = 18 markers

6. Troy bought a package of pencils ~~to share~~. There were 13 red, 7 blue, ~~and 3 green pencils~~ in the package. How many more red than blue pencils were in the package?
   13 − 7 = 6 more red pencils

Brainwork! Write a subtraction word problem. Use extra information. Give your problem to a friend to solve.

Answer Key

Page 15 — Desk-Cleaning Day

1. John found his 2 library books. Tina found 3 times as many library books as John. ~~Tina also found 8 pencils.~~ How many library books did Tina find?
   $3 \times 2 = 6$
   Tina found __6__ library books.

2. Dirk found 12 erasers, 19 pencils, ~~and 17 Halloween stickers~~. How many fewer erasers than pencils did Dirk find?
   $19 - 12 = 7$
   Dirk found __7__ fewer erasers.

3. Chad found $.45 under his math book and $.38 in his crayon box. ~~He found 11 markers and 2 pens.~~ How much money did Chad find altogether?
   $\$.45 + \$.38 = \$.83$
   Chad found __$.83__.

4. Julie cleaned her desk in 27 minutes. Christy cleaned her desk in 9 minutes. How much less time did it take Christy to clean her desk than Julie?
   $27 - 9 = 18$
   Christy took __18__ minutes less.

5. Mona found her sticker book! She also found 18 dinosaur stickers, 29 dog stickers, ~~and 6 flower stickers.~~ How many animal stickers did Mona find?
   $18 + 29 = 47$
   Mona found __47__ animal stickers.

6. Tasha found ~~5 centimeter rulers,~~ 2 quarters, and some dimes. There were 4 dimes as many dimes as quarters. How many dimes did Tasha find in her desk?
   $4 \times 2 = 8$
   Tasha found __8__ dimes.

Brainwork! Josie and Vicki each have less than 24 but more than 18 markers. Together they have 38 markers and 48 crayons. How many markers do each of the girls have?

Page 18 — Think December

Calendar for December:
Sun.	Mon.	Tues.	Wed.	Thur.	Fri.	Sat.
			1	2	3
4	5	6	7	8	9	10
11	12	13	14	15	16	17
18	19	20	21	22	23	24
25	26	27	28	29	30	31

1. If today is Monday, what day of the week will it be eight days from now? __Tuesday__
2. December 1 is on which day of the week? __Thursday__
3. What is the date of the third Tuesday in the month? __December 20__
4. Nathan has drum lessons every Wednesday. How many lessons will he have in December? __4__
5. It snowed 9 days in December. How many days did it not snow? __22__
6. Michelle can go shopping on any Friday in December. But her mother can go with her only on December 5, 14, 16, or 20. What day should Michelle plan to go shopping with her mother? __December 16__
7. Vacation begins on December 23. Annie's birthday is four days earlier. What is the date of Annie's birthday? __December 19__
8. January 1 will be on what day of the week? __Sunday__

Brainwork! Use the calendar above to find this date. It is on a weekend. It is an even number and has two digits. The two digits add up to six.

Page 14 — At the Movies

1. Teresa bought a drink for $.45 and peanuts for $.35. Later she bought popcorn for $.55. How much money did Teresa spend altogether?
   $\$.45 + \$.35 + \$.80$ $\$.80 + \$.55 = \$1.35$
   Teresa spent $1.35.

2. At the refreshment stand, Carlos saw 4 shelves with 5 boxes of popcorn on each shelf. A man bought 3 of the boxes. How many boxes were left?
   $4 \times 5 = 20$ $20 - 3 = 17$
   There were __17__ boxes left.

3. There were 34 people in Teresa's row. 9 of them were fathers and 18 were children. The rest were mothers. How many fathers were in the row?
   $34 - 9 = 25$ $25 - 18 = 7$
   or
   $9 + 18 = 27$ $34 - 27 = 7$
   There were __7__ fathers.

4. Carlos bought a large popcorn for 70¢. He paid with a 50¢ piece and a quarter. How much change did Carlos receive?
   $50¢ + 25¢ = 75¢$ $75¢ - 70¢ = 5¢$
   Carlos received __5¢__ change.

5. The theater can seat 292 people. One day 197 children and 88 adults came to see the show. How many theater seats were empty?
   $292 - 197 = 95$ $95 - 88 = 7$
   or
   $197 + 88 = 285$ $292 - 285 = 7$
   __7__ seats were empty.

6. During intermission Dad bought two boxes of popcorn for $.55 each. How much change did he receive if he gave the clerk $2.00?
   $\$2.00 - \$.55 = \$1.10$ $\$1.10 - \$.55 = \$.90$
   or
   $\$.55 + \$.55 = \$1.10$ $\$2.00 - \$1.10 = \$.90$
   Dad received __$.90__ change.

Brainwork! Write a two-step problem for a friend to solve.

Page 17 — Speeds of Animals

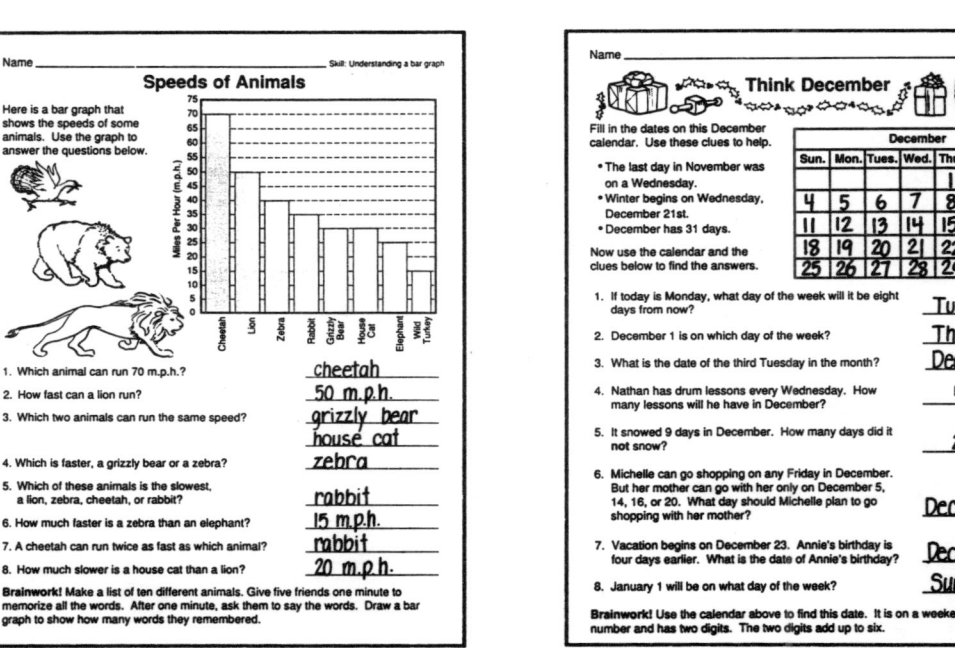

1. Which animal can run 70 m.p.h.? __cheetah__
2. How fast can a lion run? __50 m.p.h.__
3. Which two animals can run the same speed? __grizzly bear, house cat__
4. Which is faster, a grizzly bear or a zebra? __zebra__
5. Which of these animals is the slowest, a lion, zebra, cheetah, or rabbit? __rabbit__
6. How much faster can a zebra than an elephant? __15 m.p.h.__
7. A cheetah can run twice as fast as which animal? __rabbit__
8. How much slower is a house cat than a lion? __20 m.p.h.__

Brainwork! Make a list of ten different animals. Give five friends one minute to memorize all the words. After one minute, ask them to say the words. Draw a bar graph to show how many words they remembered.

Page 13 — Fun on the Playground

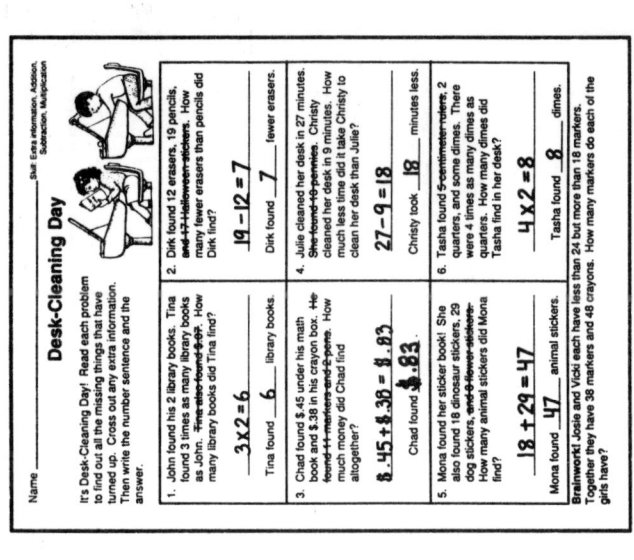

1. In the first inning of the kickball game Emily's team scored 9 runs. They scored 5 more runs the next inning and 3 in the last. How many runs did Emily's team score in all?
   $9 + 5 = 14$ $14 + 3 = 17$
   __17__ runs

2. There were 17 children running a race. 8 went to get a drink. Then 6 went to join the kickball game. How many were left in the running race?
   $17 - 8 = 9$ $17 - 14 = 3$
   or
   $9 - 6 = 3$
   __3__ children

3. There were 6 children jumping rope. 9 more joined them. The line was too long so 7 went to go swing. How many children were left jumping rope?
   $6 + 9 = 15$ $15 - 7 = 8$
   or
   $14 - 7 = 7$ $7 + 6 = 13$
   __8__ children __13__ children

4. There were 14 children swinging. 7 jumped off to go to the slide. Soon 6 more children came to swing. How many children are now swinging?

5. Linda climbed 13 steps up the ladder. She climbed down 4 steps to talk to Tara. Then she climbed up 8 steps to get to the top. How many steps high is the ladder?
   $13 - 4 = 9$ $9 + 8 = 17$
   __17__ steps high

6. There were 16 balls on the playground. 5 of the balls were basketballs and 4 were kickballs. All of the children came to play. How many kickballs were there?
   $5 + 4 = 9$ $16 - 9 = 7$
   or
   $16 - 5 = 11$ $11 - 4 = 7$
   __7__ kickballs

Brainwork! There were 17 children sliding. Six were on the ladder and 2 were going down the slide. The rest were waiting to climb the ladder. How many children were waiting to climb?

Page 16 — In the Gym

1. 18 children lined up in 3 rows to do situps. How many children were in each row?
   $18 \div 3 = 6$
   __6__ children were in each row.

2. Ms. Simms divided 36 children into 4 teams. How many children were on each team?
   $36 \div 4 = 9$
   __9__ children were on each team.

3. There are 6 tumbling mats. Ms. Simms divided 24 children into equal groups for each mat. How many children were in each group?
   $24 \div 6 = 4$
   __4__ children were in each group.

4. 16 girls want to shoot baskets. Mr. Young will have the same number of girls at 8 baskets. How many girls will be at each basket?
   $16 \div 8 = 2$
   __2__ girls will be at each basket.

5. 3 equal groups of children want to run relay races. There are 21 children. How many children are in each group?
   $21 \div 3 = 7$
   __7__ children are in each group.

6. 32 boys lined up in 4 equal rows to do jumping jacks. How many boys were in each row?
   $32 \div 4 = 8$
   __8__ boys were in each row.

7. At the end of class, 27 kickballs must be put equally into 3 bags. How many kickballs fit in each bag?
   $27 \div 3 = 9$
   __9__ kickballs fit in each bag.

Brainwork! Tom's dog eats 3 bones each day. How long will a box of 18 bones last? Draw pictures to prove your answer.

© Frank Schaffer Publications, Inc.   113   FS-32022 Math Activities

Answer Key

Page 19

Possible solutions — Skill: Logical thinking

Oodles of Tadpoles

Use the clues to find out how many tadpoles each student has.

1. Yolanda has an even number of tadpoles. If you take out 4, there will be 6 in her jar. How many tadpoles does Yolanda have?
   $10 - 4 = 6$
   10 tadpoles

2. Greg has fewer than 29 tadpoles. He has more than 25. He has an odd number. How many tadpoles does Greg have?
   29, 28, **27**, 26, 25
   27 tadpoles

3. Pablo has an odd number of tadpoles. If you double his number, he will have 18. How many tadpoles does Pablo have?
   $9 + 9 = 18$
   9 tadpoles

4. Megan has 20 tadpoles in her jar. Danny has twice as many tadpoles in his jar. How many tadpoles does Danny have?
   $20 \times 2 = 40$
   40 tadpoles

5. Kaithy has more than 7 but fewer than 15 tadpoles. If she counts them by fives, there is one left over. How many tadpoles does Kaithy have?
   5, 10, **15**
   $10 + 1 = 11$
   11 tadpoles

6. Paul has more than 50 tadpoles. He has fewer than 75. The two digits in the number are the same. The sum of the digits is 12. How many tadpoles does Paul have?
   55, 66
   $6 + 6 = 12$
   66 tadpoles

7. Todd has more than 30 but less than 40 tadpoles. You say the number when you count by twos and when you count by threes. How many tadpoles does Todd have?
   30, 32, 34, **36**, 38
   33, 34, **36**
   36 tadpoles

Brainwork! Tadpoles each have two eyes and one tail. How many tadpoles does Craig have in his jar? Hint: There are 4 more eyes than tails.

Page 19

Page 20

Computer Fun — Skill: Mixed strategies

Andrew and Gina have been playing computer games. Solve each problem. Write the answer.

1. Andrew scored 48 points in Math Magic. Gina scored 75 points. How many fewer points did Andrew score than Gina?
   $75 - 48 = 27$
   Andrew scored **27** fewer points.

2. Gina played Spelling Invaders 5 times. Her highest score was 49. Andrew's highest score was 68. How much higher was Andrew's highest score than Gina's?
   $68 - 49 = 19$
   Andrew's score was **19** points higher.

3. Gina's score in Punctuation Trouble was less than 250 but more than 125. It is an odd number. The three digits are 2, 5, and 2. What was Gina's score?
   Gina's score was **225**.

4. In Fraction Frogger Andrew had 60 points. Then he lost 35 points. Finally he scored 55 more points for getting Frog across the river. What was Paul's final score?
   $60 - 35 = 25$
   $25 + 55 = 80$
   Paul's final score was **80**.

5. Gina saw a pattern in her Amazing Reader scores. She scored 68, 78, 88, 98, 108, and 118. Explain the pattern she saw.
   She scored 10 more points each time she played.

6. Andrew played Math Magic 3 times. Gina played it 8 times as often as Andrew. How many games of Math Magic did Gina play? Circle the letter of the correct number sentence and answer.
   a. $3 + 8 = 11$ games
   b. $8 \times 3 = 24$ games
   c. $8 \times 3 = 24$ points

Brainwork! Write two word problems about computer games. Ask a friend to solve them.

Page 20

Page 21

Mystery picture: blue and yellow kangaroo

Page 22

Mystery picture: green dolphin looking out of ship portal

Page 23

Mystery picture: white bird flying over a grove of green trees and a blue waterfall

Page 24

Mystery picture: red, yellow, orange, and blue truck in front of houses

Answer Key

Page 25
Mystery picture: brown and yellow rabbit hopping over flowers

Page 26
Mystery picture: two people driving a yellow, red, orange, and black car

Page 27
Mystery picture; brown, orange, and yellow buffalo

Page 28
Mystery picture; brown and yellow dinosaur with one orange spot

Page 29
Mystery picture: puppet wearing an orange and blue costume holding a black stick

Page 30
Mystery picture: black and yellow skunk smelling an orange flower

115

© Frank Schaffer Publications, Inc.

FS-32022 Math Activities

Answer Key

Page 31
Mystery picture: yellow and green octupus and an orange squid

Page 32
Mystery picture: orange and brown donkey and a yellow duck

Page 33
Mystery picture: predominantly yellow car taking a shower

Page 34
Mystery picture: clown with yellow face, green nose, and red hair eating an ice-cream cone

Page 35
Mystery picture: reclining black and yellow cat wearing red boots

Page 36
Mystery picture: brown moose standing in a grove of green trees

Answer Key

Mystery picture: boy wearing an orange shirt, blue pants, and black cap swinging a yellow baseball bat

Page 38

Mystery picture: toy soldier with yellow hair, black hat and gloves, and red and blue costume

Page 39

Mystery picture: orange and yellow flamingo standing in front of a blue sky

Page 37

Mystery picture: red silhouette of a girl riding a yellow and blue merry-go-round horse

Page 40

Mystery picture: brown horse with an orange mane hiding behind green leaves

Page 41

Mystery picture: flying black bird near a black eagle sitting on branch

Page 42

© Frank Schaffer Publications, Inc.

117

FS-32022 Math Activities

Answer Key

Page 43
Mystery picture: gnome wearing a green costume and red cap standing near a brown mushroom

Page 44
Mystery picture: red car with blue trim in front of green trees and yellow city buildings

Page 45
Mystery picture: silhouettes of a yellow rabbit and an orange camel

Page 46
Mystery picture: boy wearing a red and yellow cap, orange shirt, and blue pants

Page 47
Mystery picture: flying red bird and black song notes near an orange lion wearing a red crown

Page 48
Mystery picture: silhouettes of a red elephant and a black horse

© Frank Schaffer Publications, Inc. 118 FS-32022 Math Activities

Answer Key

Mystery picture: marching band leader with a yellow baton wearing a red costume and yellow boots
Page 51

Mystery picture: clown marionette wearing a green jacket and pants, yellow and orange shirt, and red bow tie
Page 54

Mystery picture: jester wearing a red hat and red shoes playing an orange guitar
Page 50

Mystery picture: brown and yellow lion
Page 53

Mystery picture: orange and yellow flamingo standing next to an orange and yellow fish
Page 49

Mystery picture: brown squirrel with yellow face and chest holding an orange book
Page 52

Answer Key

Mystery picture:
red car
with yellow top
and blue wheels

Page 57

Mystery picture:
orange lemur perched
on a brown branch
near green trees

Page 60

Mystery picture:
yellow flying horse
with orange mane
and multicolored wings

Page 56

Mystery picture:
painter wearing
blue coveralls and
yellow shirt standing
next to a blue can
of spilled red paint

Page 59

Mystery picture:
man wearing green
shirt and pants hanging a
picture with a red frame

Page 55

Mystery picture:
yellow pilot in a red,
brown and orange
airplane

Page 58

© Frank Schaffer Publications, Inc.

FS-32022 Math Activities

Answer Key

Mystery picture: yellow, brown, and orange giraffe standing next to a brown branch with yellow leaves Page 61	**Mystery picture:** orange and yellow dinosaurs eating a green plant Page 65
Mystery picture: flying lion with yellow wings and tail near a flying brown bird Page 62	**Mystery picture:** chef wearing a blue and yellow costume Page 66
Mystery picture: red and blue jet Page 63	**Mystery picture:** cowboy wearing a red shirt and blue pants riding a bucking yellow and brown horse Page 67
Mystery picture: brown and orange bird with multicolored tail Page 64	**Mystery picture:** brown and red car with yellow lights, bumper, and hub caps Page 68

Answer Key

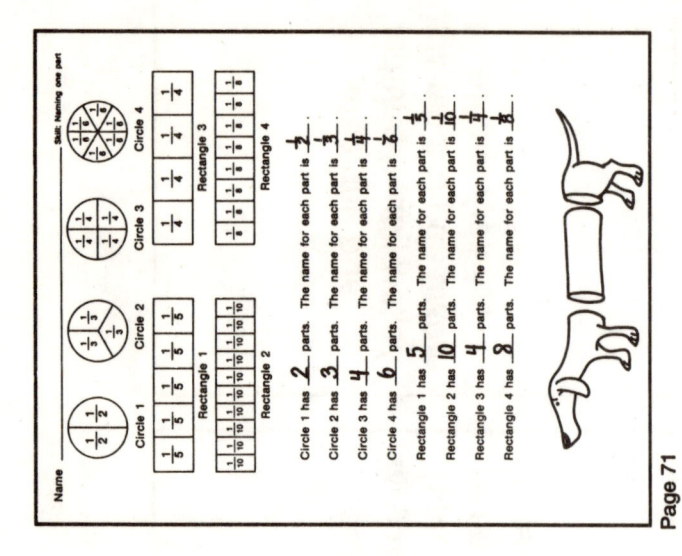

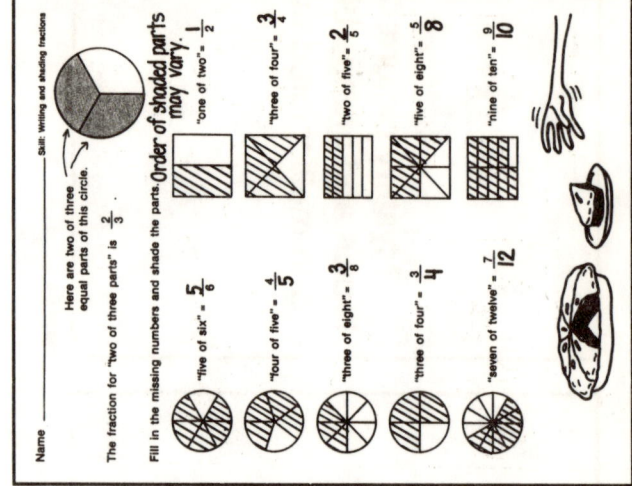

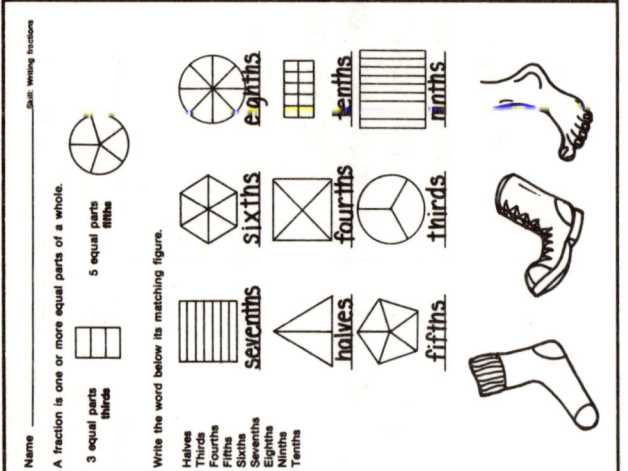

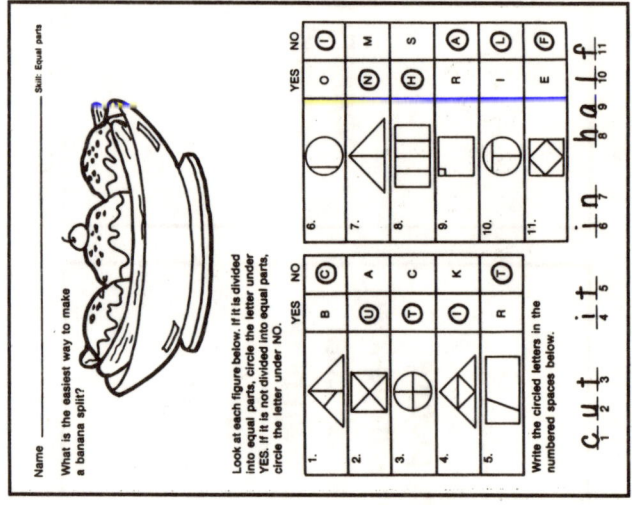

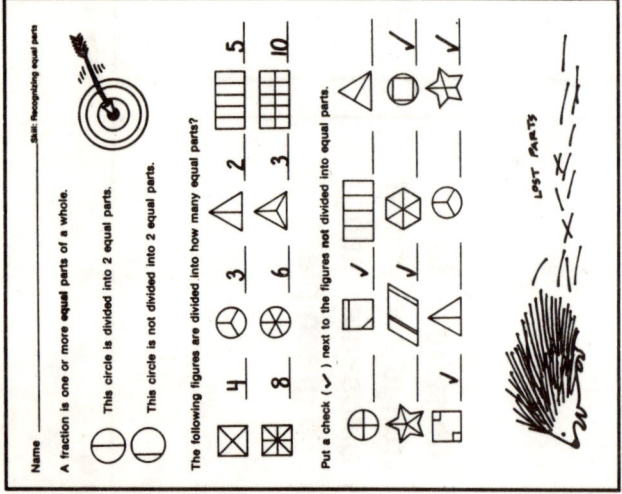

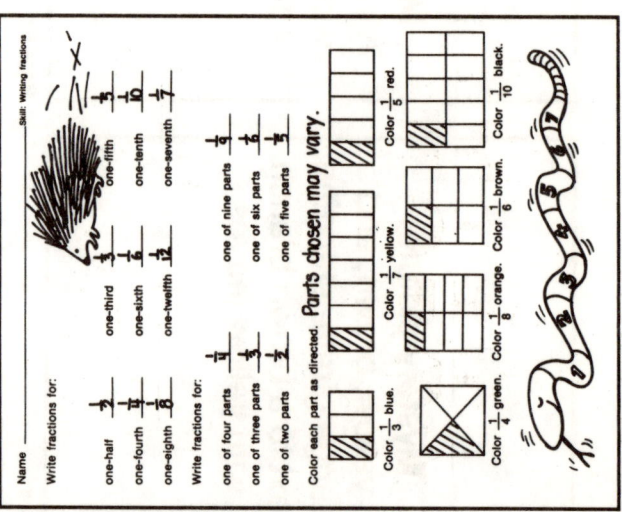

Page 122

Answer Key

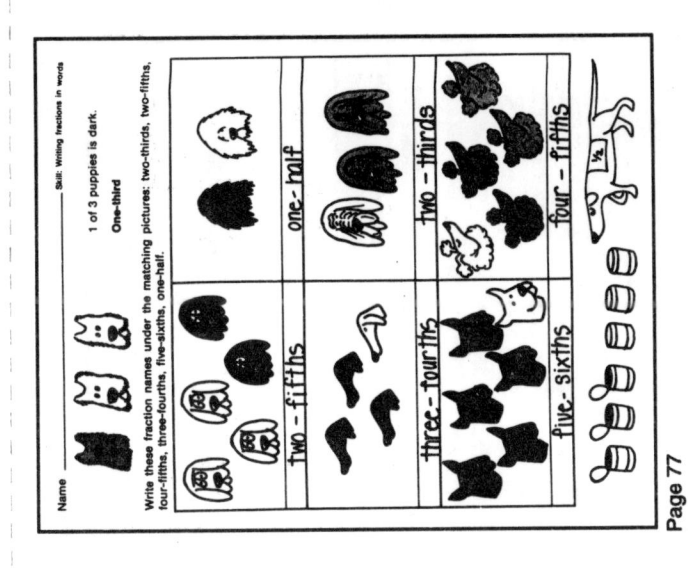

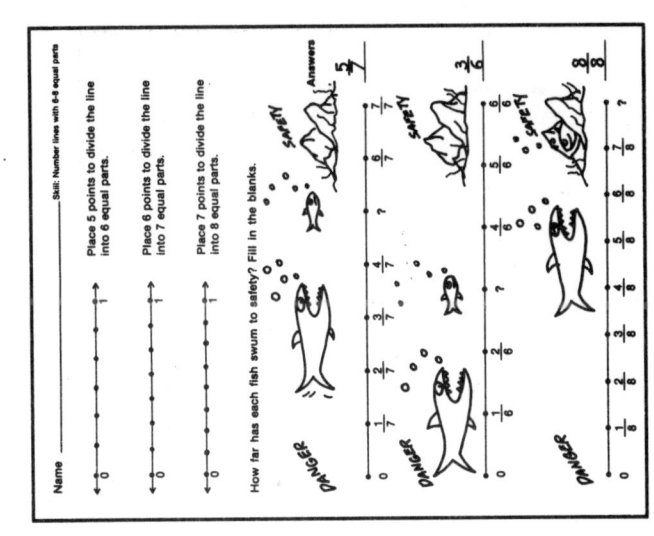

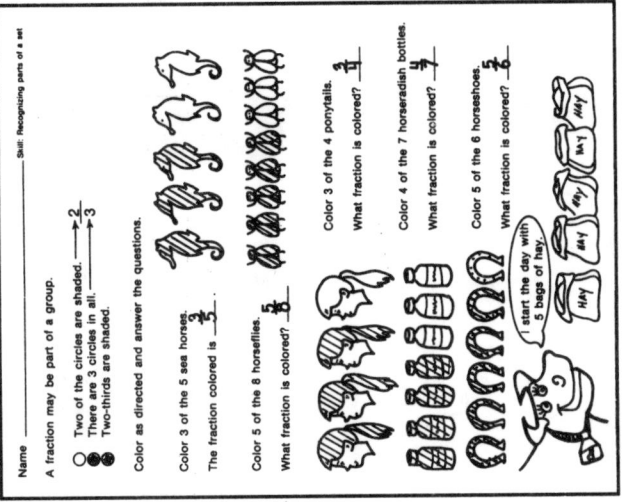

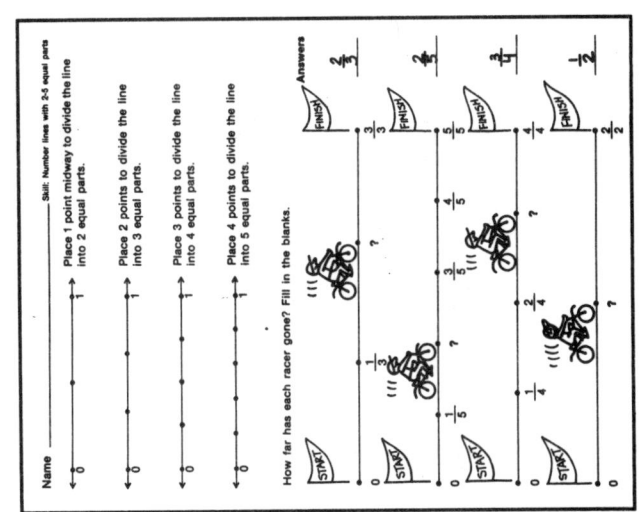

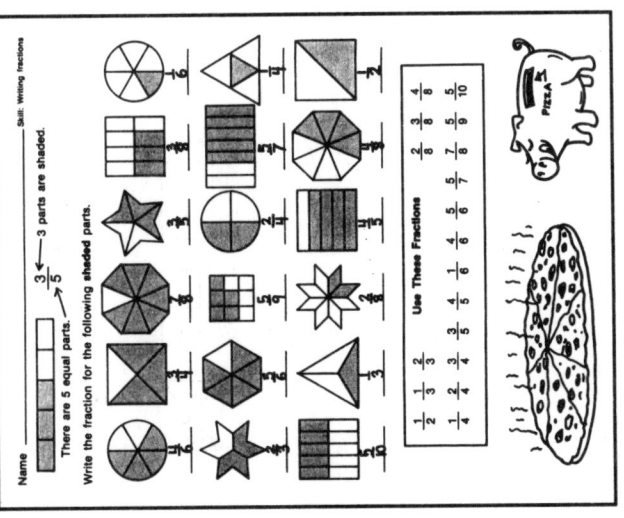

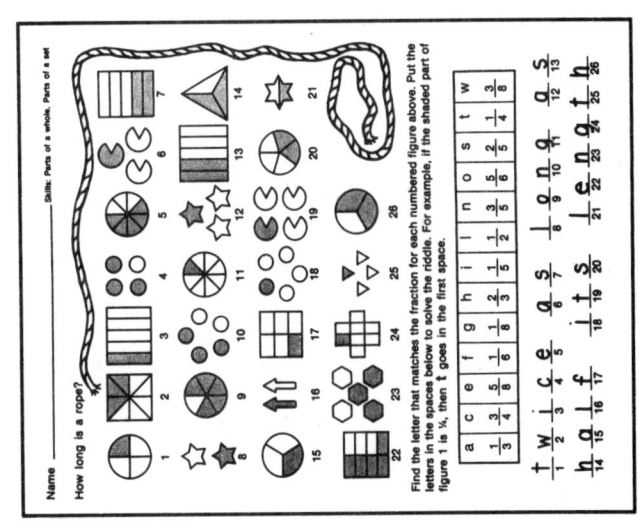

Answer Key

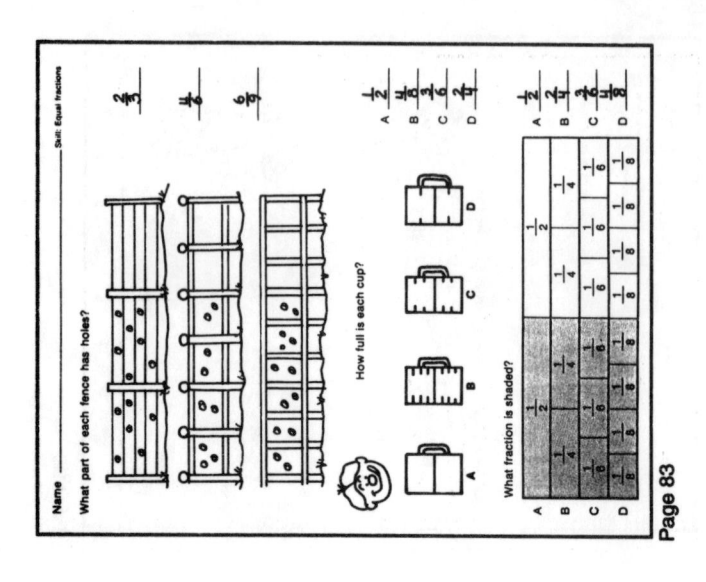

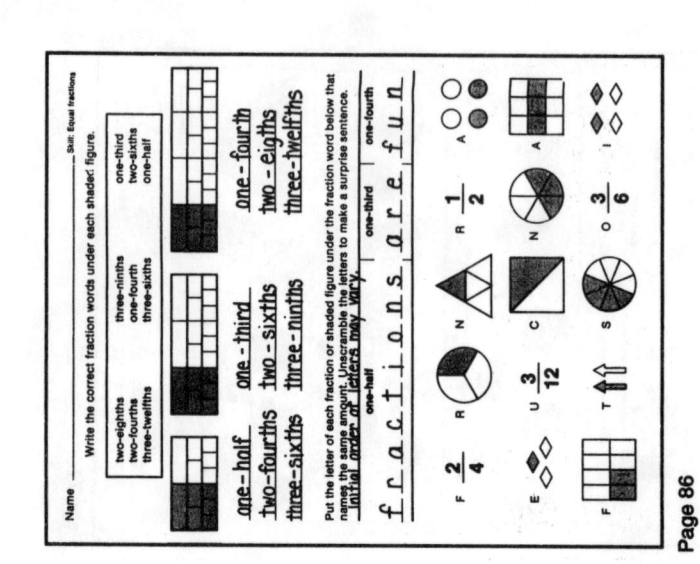

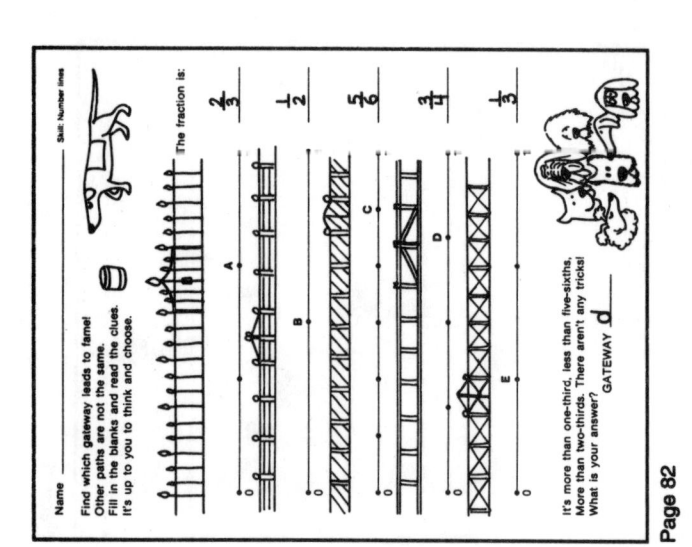

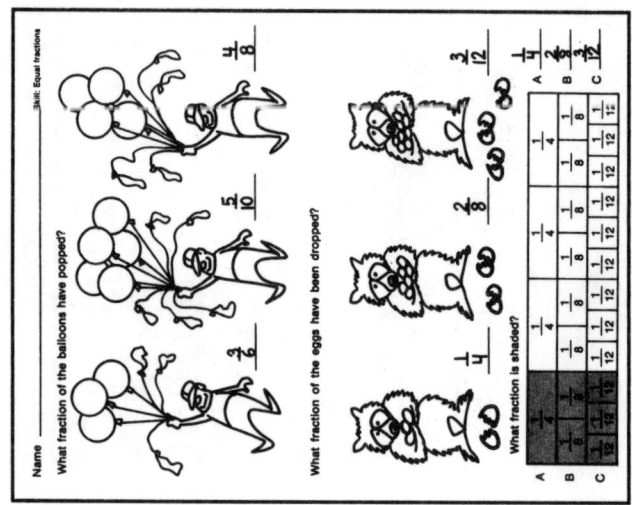

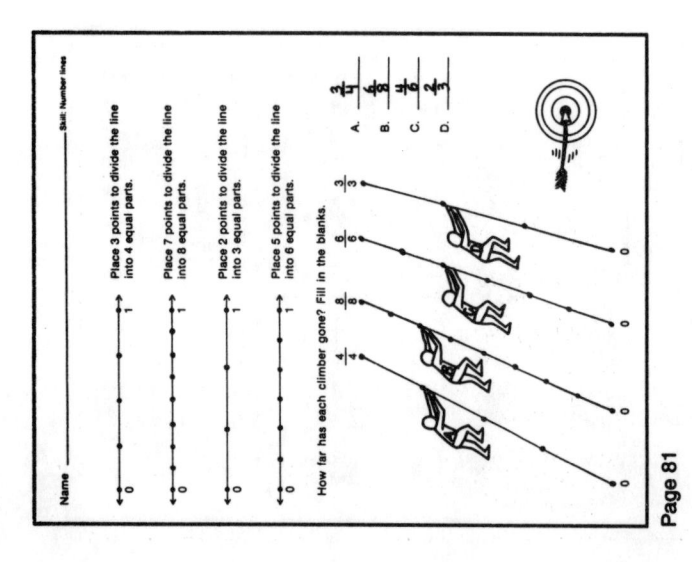

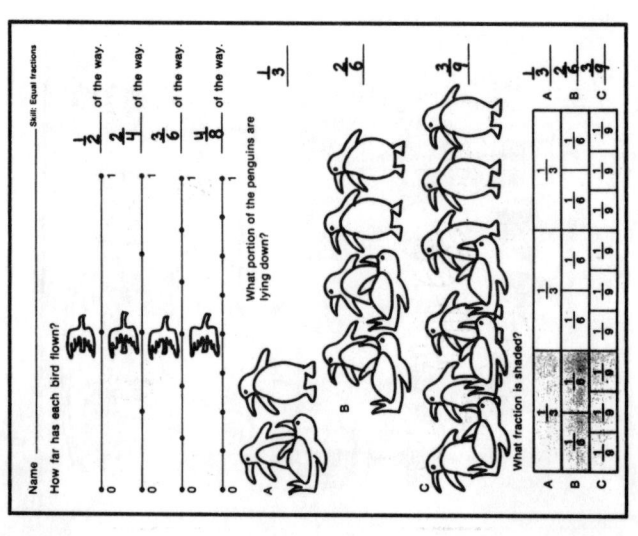

Answer Key

Page 89
Skill: Comparisons (3 digit)

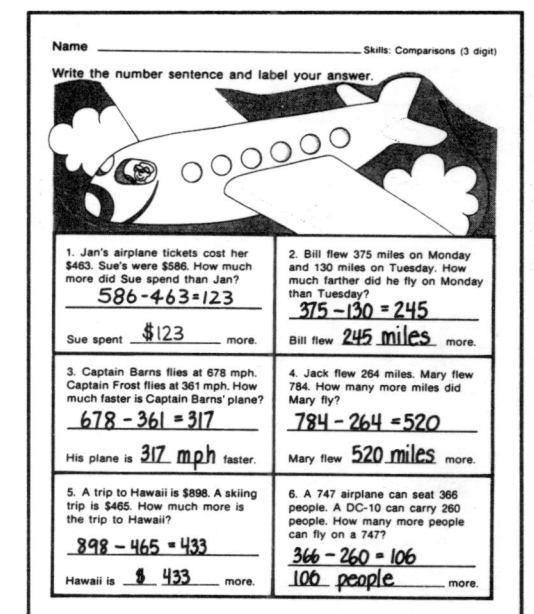

1. Jan's airplane tickets cost her $463. Sue's were $586. How much more did Sue spend than Jan?
586 − 463 = 123
Sue spent $123 more.

2. Bill flew 375 miles on Monday and 130 miles on Tuesday. How much farther did he fly on Monday than Tuesday?
375 − 130 = 245
Bill flew 245 miles more.

3. Captain Barns flies at 678 mph. Captain Frost flies at 361 mph. How much faster is Captain Barns' plane?
678 − 361 = 317
His plane is 317 mph faster.

4. Jack flew 264 miles. Mary flew 784. How many more miles did Mary fly?
784 − 264 = 520
Mary flew 520 miles more.

5. A trip to Hawaii is $898. A skiing trip is $465. How much more is the trip to Hawaii?
898 − 465 = 433
Hawaii is $433 more.

6. A 747 airplane can seat 366 people. A DC-10 can carry 260 people. How many more people can fly on a 747?
366 − 260 = 106
106 people more.

Page 92
Skill: 3 digit regrouping—comparisons

hippo	924 pounds
bear	826 pounds
gorilla	342 pounds
zebra	325 pounds
lion	267 pounds
seal	129 pounds
wolf	102 pounds

1. How much more does the hippo weigh than the lion?
924 − 267 = 657
657 pounds more.

2. How much less does the wolf weigh than the gorilla?
342 − 102 = 240
240 pounds less.

3. How much difference in weight is there between the zebra and the seal?
325 − 129 = 196
The difference is 196 pounds.

4. How much difference in weight is there between the bear and the gorilla?
826 − 342 = 484
The difference is 484 pounds.

5. How much less does the bear weigh than the hippo?
924 − 826 = 98
98 pounds less.

6. How much more does the gorilla weigh than the lion?
342 − 267 = 75
75 pounds.

Page 88
Skill: 3 digit addition

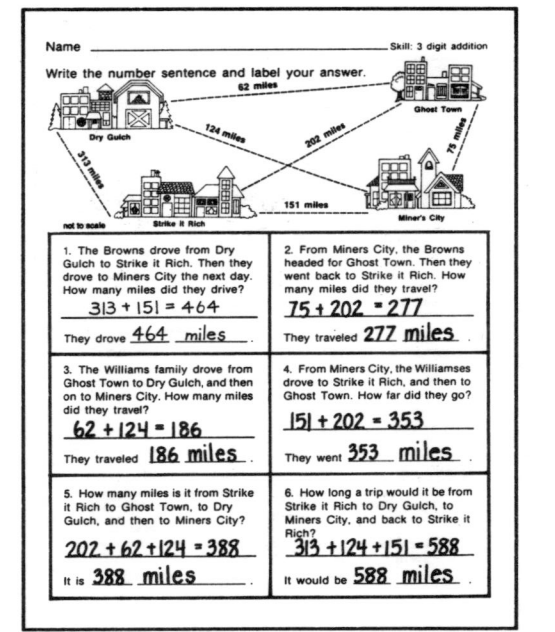

1. The Browns drove from Dry Gulch to Strike it Rich. Then they drove to Miners City the next day. How many miles did they drive?
313 + 151 = 464
They drove 464 miles.

2. From Miners City, the Browns headed for Ghost Town. Then they went back to Strike it Rich. How many miles did they travel?
75 + 202 = 277
They traveled 277 miles.

3. The Williams family drove from Ghost Town to Dry Gulch, and then on to Miners City. How many miles did they travel?
62 + 124 = 186
They traveled 186 miles.

4. From Miners City, the Williames drove to Strike it Rich, and then to Ghost Town. How far did they go?
151 + 202 = 353
They went 353 miles.

5. How many miles is it from Strike it Rich to Ghost Town, to Dry Gulch, and then to Miners City?
202 + 62 + 124 = 388
It is 388 miles.

6. How long a trip would it be from Strike it Rich to Dry Gulch, to Miners City, and back to Strike it Rich?
313 + 124 + 151 = 588
It would be 588 miles.

Page 91
Skill: 3 digit addition—regrouping

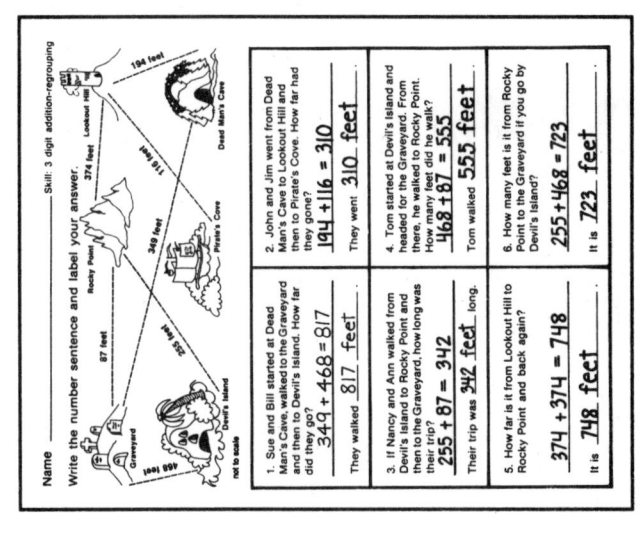

1. Sue and Bill started at Dead Man's Cave and then to the Graveyard and then to Devil's Island. How far did they go?
349 + 468 = 817
They walked 817 feet.

2. John and Jim went from Dead Man's Cave to Lookout Hill and then to Pirate's Cove. How far had they gone?
194 + 116 = 310
They went 310 feet.

3. If Nancy and Ann walked from Devil's Island to Rocky Point and then to the Graveyard, how long was their trip?
255 + 87 = 342
Their trip was 342 feet long.

4. Tom started at Devil's Island and headed for the Graveyard. From there, he walked to Rocky Point. How many feet did he walk?
468 + 87 = 555
Tom walked 555 feet.

5. How far is it from Lookout Hill to Rocky Point and back again?
374 + 374 = 748
It is 748 feet.

6. How many feet is it from Rocky Point to the Graveyard if you go by Devil's Island?
255 + 468 = 723
It is 723 feet.

Page 87
Skill: Recognizing equal fractions

Color red all balloons that show one-half.
Color green all balloons that show one-third.
Color blue all balloons that show one-fourth.

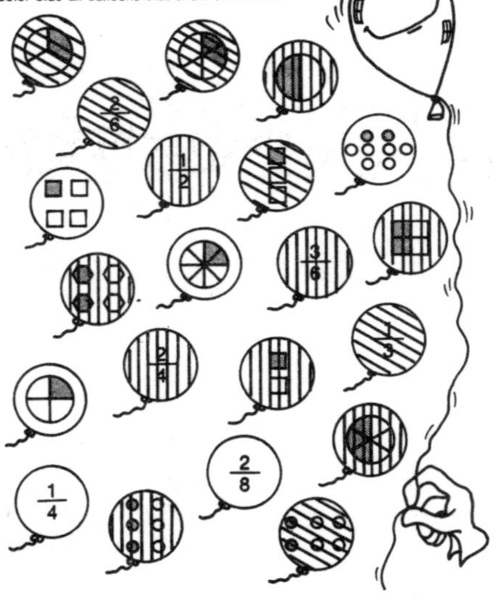

Page 90
Skill: 3 digit addition and subtraction

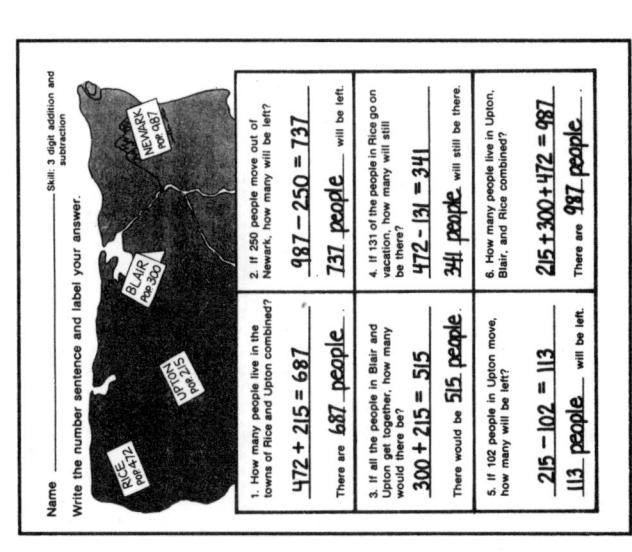

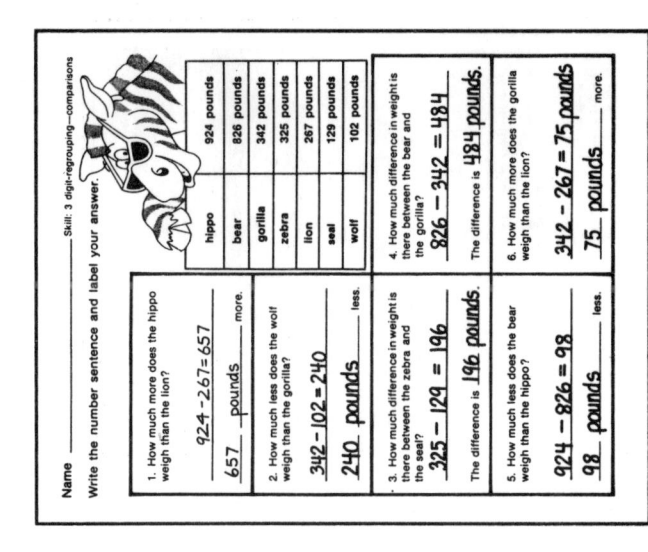

1. How many people live in the towns of Rice and Upton combined?
472 + 215 = 687
There are 687 people.

2. If 250 people move out of Newark, how many will be left?
987 − 250 = 737
737 people will be left.

3. If all the people in Blair and Upton get together, how many would there be?
300 + 215 = 515
There would be 515 people.

4. If 131 of the people in Rice go on vacation, how many will still be there?
472 − 131 = 341
341 people will still be there.

5. If 102 people in Upton move, how many will be left?
215 − 102 = 113
113 people will be left.

6. How many people live in Upton, Blair, and Rice combined?
215 + 300 + 472 = 987
There are 987 people.

Answer Key

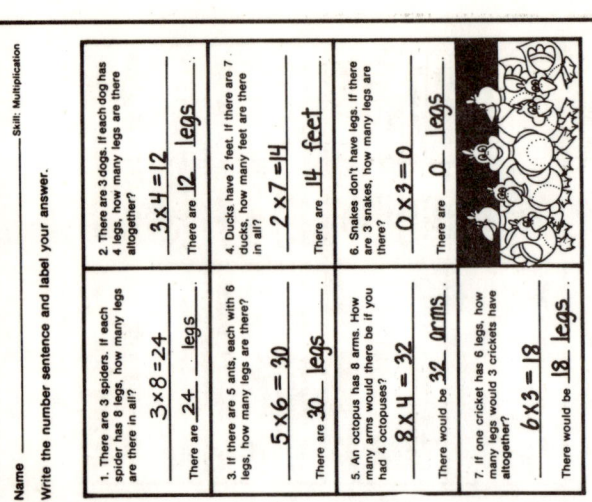

Page 95
Skill: Multiplication

1. There are 3 spiders. If each spider has 8 legs, how many legs are there in all?
   $3 \times 8 = 24$
   There are 24 legs.

2. There are 3 dogs. If each dog has 4 legs, how many legs are there altogether?
   $3 \times 4 = 12$
   There are 12 legs.

3. If there are 5 ants, each with 6 legs, how many legs are there?
   $5 \times 6 = 30$
   There are 30 legs.

4. Ducks have 2 feet. If there are 7 ducks, how many feet are there in all?
   $2 \times 7 = 14$
   There are 14 feet.

5. An octopus has 8 arms. How many arms would there be if you had 4 octopuses?
   $8 \times 4 = 32$
   There would be 32 arms.

6. Snakes don't have legs. If there are 3 snakes, how many legs are there?
   $0 \times 3 = 0$
   There are 0 legs.

7. If one cricket has 6 legs, how many legs would 3 crickets have altogether?
   $6 \times 3 = 18$
   There would be 18 legs.

Page 98
Skill: Multiplication or division

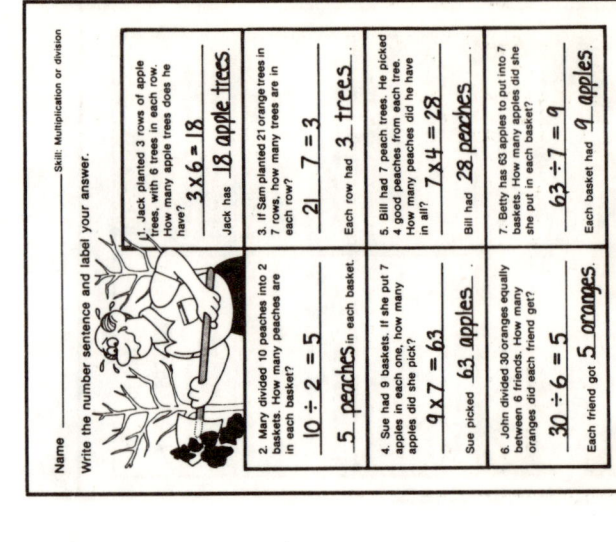

1. Jack planted 3 rows of apple trees, with 6 trees in each row. How many apple trees does he have?
   $3 \times 6 = 18$
   Jack has 18 apple trees.

2. If Sam planted 21 orange trees in 7 rows, how many trees are in each row?
   $21 \div 7 = 3$
   Each row had 3 trees.

3. Bill had 7 peach trees. He picked 4 good peaches from each tree. How many peaches did he have in all?
   $7 \times 4 = 28$
   Bill had 28 peaches.

4. Betty has 63 apples to put into 7 baskets. How many apples did she put in each basket?
   $63 \div 7 = 9$
   Each basket had 9 apples.

5. Mary divided 10 peaches into 2 baskets. How many peaches are in each basket?
   $10 \div 2 = 5$
   5 peaches in each basket.

6. Sue had 9 baskets. If she put 7 apples in each one, how many apples did she have?
   $9 \times 7 = 63$
   Sue picked 63 apples.

7. John divided 30 oranges equally between 6 friends. How many oranges did each friend get?
   $30 \div 6 = 5$
   Each friend got 5 oranges.

Page 94
Skill: Addition and Subtraction—3+ Digits with Regrouping

How Inventive are Inventors?
Write an equation and label your answer.

1. Professor Twitcher invented 168 wacky ways to wake a person up and 303 wacky ways to put them to sleep. How many wacky ways did he invent all together?
   $168 + 303 = 471$
   471 wacky ways

2. Dr. Dunderhead lost 158 inventions in his lab when it exploded. He had 648 inventions in the lab. How many were saved?
   $648 - 158 = 490$
   490 inventions

3. The safety pin was invented in 1849. The zipper was invented 42 years later. In what year was the zipper invented?
   $1849 + 42 = 1891$
   1891

4. Edison patented more than 1100 inventions in his lifetime. He lived from 1847-1931. How old was he when he died?
   $1931 - 1847 = 84$
   84 years old

5. 400 inventors entered a contest to make jet-powered canoes. 218 canoes stayed afloat. How many sank?
   $400 - 218 = 182$
   182 canoes

6. Super Smart Sally sold her design for a singing swing for $2,500. Her sister Sue sold her design for silent skates for $1,800. How much did they earn all together?
   $\$2,500.00 + \$1,800.00 = \$4,300.00$
   $\$4,300.00

Page 97
Skill: Division

1. There were 16 boy scouts. They slept in 8 tents. How many boys were in each tent?
   $16 \div 8 = 2$
   Each tent had 2 boy scouts.

2. If 32 girl scouts stayed in 4 cabins, how many were in each cabin?
   $32 \div 4 = 8$
   Each cabin had 8 girl scouts.

3. There are 6 large tables in the mess hall. If there are 54 scouts, how many will sit at each table?
   $54 \div 6 = 9$
   9 scouts at each table.

4. 63 girl scouts wanted to go boating. If there are 9 boats, how many girls will be in each boat?
   $63 \div 9 = 7$
   7 girl scouts in each boat.

5. Seven groups of boy scouts went hiking. There were 42 boys. How many were in each group?
   $42 \div 7 = 6$
   Each group had 6 boy scouts.

6. If 25 scouts make 5 campfires, how many scouts will sit at each campfire?
   $25 \div 5 = 5$
   5 scouts at each campfire.

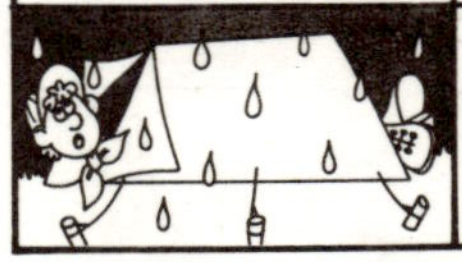

7. The arts and crafts leader can take groups of 9 scouts at a time. If 27 scouts sign up, how many groups would come?
   $27 \div 9 = 3$
   There would be 3 groups.

Page 93
Skill: Addition and Subtraction—2-3 Digits with Regrouping

Sports of all Sorts!
Write an equation and label your answer.

1. Stan Superstar did 210 situps in 3 minutes. Allie Ace did 185. How many more situps did Stan S. do than Allie A.?
   $210 - 185 = 25$
   25 situps

2. In last night's basketball game, Marty made 23 points. Mary made 19. Millie and Missie each made 15. Susie made 3. How many points did they make all together?
   $23 + 19 + 15 + 15 + 3 = 75$
   75 points

3. 609 soccer fans came to see the big game. 352 sat on one side. How many soccer fans sat on the other side of the field?
   $609 - 352 = 257$
   257 soccer fans

4. Baron Bounder hit the baseball 132 feet into center field. If the center fielder was standing 95 feet from home plate, how far did he have to run to get the ball?
   $132 - 95 = 37$
   37 feet

5. Sara Sidesaddle rode her pinto pony, Peanuts, 245 yards on Trail A, then 145 yards on Trail B. How far did Sara and Peanuts go?
   $245 + 145 = 390$
   390 yards

6. Gary Gutter and Sam Splitz bowled two games at Trendy Ten Pins. S.S. scored 97 and 132. G.G. scored 102 and 115. How many more pins did S.S. knock down?
   $97 + 132 = 229$
   $102 + 115 = 217$
   $229 - 217 = 12$
   12 pins

Page 96
Skill: Multiplication

Write the number sentence and label your answer.

1. A football team gets 6 points for a touchdown. How many points would 8 touchdowns be?
   $6 \times 8 = 48$
   It would be 48 points.

2. Baseball games have 9 innings. If each inning has 3 outs, how many outs are in a game?
   $9 \times 3 = 27$
   There are 27 outs.

3. If you swam 8 laps a day for 7 days, how many laps would you swim altogether?
   $8 \times 7 = 56$
   That would be 56 laps.

4. There are 4 people at each ping pong table. If there are 9 tables, how many people are playing?
   $4 \times 9 = 36$
   There are 36 people.

5. Dad jogs 6 miles a day. How many miles would he jog in 4 days?
   $6 \times 4 = 24$
   Dad would jog 24 miles.

6. There are 5 players on a basketball team. If they each score 9 points, what would the team's score be?
   $5 \times 9 = 45$
   The score would be 45 points.

7. In tennis, you play 6 games each set. If you play 7 sets, how many games would you play?
   $6 \times 7 = 42$
   You would play 42 games.

Answer Key

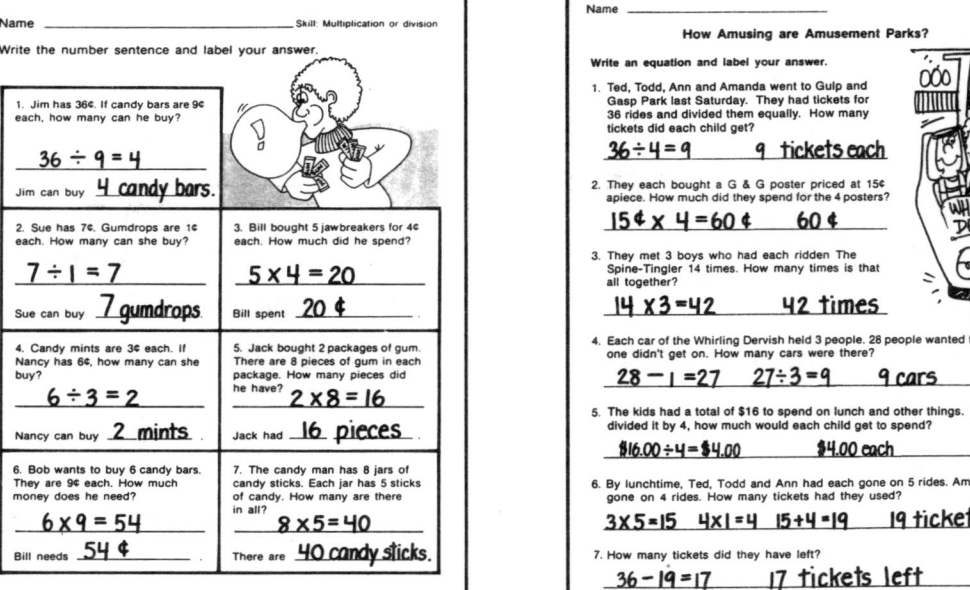

Page 99
Skill: Multiplication or division
Write the number sentence and label your answer.

1. Jim has 36¢. If candy bars are 9¢ each, how many can he buy?
   $36 \div 9 = 4$ Jim can buy 4 candy bars.
2. Sue has 7¢. Gumdrops are 1¢ each. How many can she buy?
   $7 \div 1 = 7$ Sue can buy 7 gumdrops.
3. Bill bought 5 jawbreakers for 4¢ each. How much did he spend?
   $5 \times 4 = 20$ Bill spent 20¢.
4. Candy mints are 3¢ each. If Nancy has 6¢, how many can she buy?
   $6 \div 3 = 2$ Nancy can buy 2 mints.
5. Jack bought 2 packages of gum. There are 8 pieces of gum in each package. How many pieces did he have?
   $2 \times 8 = 16$ Jack had 16 pieces.
6. Bob wants to buy 6 candy bars. They are 9¢ each. How much money does he need?
   $6 \times 9 = 54$ Bill needs 54¢.
7. The candy man has 8 jars of candy sticks. Each jar has 5 sticks of candy. How many are there in all?
   $8 \times 5 = 40$ There are 40 candy sticks.

Page 100
How Amusing are Amusement Parks?
Write an equation and label your answer.

1. Ted, Todd, Ann and Amanda went to Gulp and Gasp Park last Saturday. They had tickets for 36 rides and divided them equally. How many tickets did each child get?
   $36 \div 4 = 9$ 9 tickets each
2. They each bought a G & G poster priced at 15¢ apiece. How much did they spend for the 4 posters?
   $15¢ \times 4 = 60¢$ 60¢
3. They met 3 boys who had ridden The Spine-Tingler 14 times. How many times is that all together?
   $14 \times 3 = 42$ 42 times
4. Each car of the Whirling Dervish held 3 people. 28 people wanted to ride. Only one didn't get on. How many cars were there?
   $28 - 1 = 27$ $27 \div 3 = 9$ 9 cars
5. The kids had a total of $16 to spend on lunch and other things. If they divided it by 4, how much would each child get to spend?
   $16.00 \div 4 = 4.00 $4.00
6. By lunchtime, Ted, Todd and Ann had each gone on 5 rides. Amanda had gone on 4 rides. How many tickets had they used?
   $3 \times 5 = 15$ $4 \times 1 = 4$ $15 + 4 = 19$ 19 tickets
7. How many tickets did they have left?
   $36 - 19 = 17$ 17 tickets left

Page 101
I Love a Parade!
Write an equation and label your answer.

1. The Turkey Bowl Parade will have 8 bands. If there are 10 in each band, how many band members will be marching?
   $8 \times 10 = 80$ 80 band members
2. There will be 9 rows to sit in along the way. Each row can hold 9 people. How many people can sit in all nine rows?
   $9 \times 9 = 81$ 81 people
3. 110 people are coming to see the parade. How many will not be able to sit in the rows?
   $110 - 81 = 29$ 29 people
4. The drill team from Tempo High has 29 girls. One is a baton twirler. The rest will march in 4 rows. How many girls will be in each row?
   $29 - 1 = 28$ $28 \div 4 = 7$ 7 girls
5. How many rows of horses can be lined up 8 across if there are 56 horses?
   $56 \div 8 = 7$ 7 rows of horses
6. If there are 9 floats each carrying 8 turkeys and 7 floats each carrying 9 ducks, which floats are carrying more birds: the turkey float or the duck float? How many more?
   $9 \times 8 = 72$ turkeys $7 \times 9 = 63$ ducks
   $72 - 63 = 9$ 9 more turkeys

Page 102
From Little Wheels to Big Wheels
Write an equation and label your answer.

1. If one pair of roller skates costs $36, how much do 4 pairs cost?
   $36.00 \times 4 = 144.00 $144.00
2. Dizzy Turner did 5 "360's" in a row on his skateboard. If there are 360 degrees in a circle, how many degrees did he turn all together?
   $360 \times 5 = 1800$ 1800 degrees
3. It took Tony Tenspeed 2 days to travel 68 miles on his bike. If he went the same distance each day, how many miles did he pedal per day?
   $68 \div 2 = 34$ 34 miles
4. If a go-cart could go 27 miles per hour, how far could it go in 5 hours?
   $27 \times 5 = 135$ 135 miles
5. Robert rode in his Rabbit from Redfield to Greenfield, a distance of 99 miles. It took 3 hours. How fast was he driving?
   $99 \div 3 = 33$ 33 m.p.h.
6. Abe drove his eighteen-wheeler for 4 hours at 54 mph. Sid drove his sixteen-wheeler for 5 hours at 45 mph. How many more miles did Sid drive than Abe?
   $54 \times 4 = 216$ miles $45 \times 5 = 225$ miles
   $225 - 216 = 9$ 9 miles

Page 103
Glub for Gold
Write an equation and label your answer.

1. 2 deep-sea divers spent $350 each to search for sunken treasure. How much did they spend all together?
   $350.00 \times 2 = 700.00 $700.00 each
2. Freddy Fin found the hulk of a galleon in 6 hours after diving and searching for 306 feet. How many feet per hour is that?
   $306 \div 6 = 51$ 51 feet per hour
3. 8 treasure chests were found, each containing 90 coins. How many coins were found?
   $90 \times 8 = 720$ 720 coins.
4. If 7 divers, 1 tender, and 1 captain shared these coins equally, how many coins did each keep?
   $720 \div 9 = 80$ 80 coins each
5. Each coin was worth $8. If they sold all the coins, how much would each one get?
   $8.00 \times $8.00 = 640.00 $640.00 each
6. How much money did each diver earn on this trip?
   $640.00 - $350.00 = 290.00 $290.00 each

Page 104
Bunk on a Trunk
Skill: Addition and Subtraction of Like Fractions
Write an equation and label your answer.

1. The tree house was $\frac{1}{5}$ mile east of Tom's house and $\frac{2}{5}$ mile west of Tony's house. How far was it from Tom's to Tony's?
   $\frac{1}{5} + \frac{2}{5} = \frac{3}{5}$ $\frac{3}{5}$ of a mile
2. Tom cleaned the tree house $\frac{1}{6}$ of an hour on Sunday and $\frac{4}{6}$ of an hour on Monday. How long did Tom clean?
   $\frac{1}{6} + \frac{4}{6} = \frac{5}{6}$ $\frac{5}{6}$ of an hour
3. Terry tried to sneak up the rope ladder. She got $\frac{9}{10}$ of the way and slipped $\frac{6}{10}$ down. Where did that leave her?
   $\frac{9}{10} - \frac{6}{10} = \frac{3}{10}$ $\frac{3}{10}$ of the way
4. Barbie brought them $\frac{3}{4}$ of a pizza. If she ate $\frac{1}{4}$, what did that leave Tom and Tony?
   $\frac{3}{4} - \frac{1}{4} = \frac{2}{4}$ $\frac{2}{4}$ of a pizza
5. $\frac{1}{6}$ of the tree house was covered with wooden planks and $\frac{4}{6}$ was covered with a sheet. How much was covered?
   $\frac{1}{6} + \frac{4}{6} = \frac{5}{6}$ $\frac{5}{6}$ covered
6. Their gallon bucket of water held $\frac{1}{8}$ gal. of water left in it. They lowered it by rope and Barbie poured $\frac{2}{8}$ of a gal. of water into it. On the way up, $\frac{3}{8}$ gal. spilled out. How much was left?
   $\frac{1}{8} + \frac{2}{8} = \frac{3}{8}$ $\frac{6}{8} - \frac{3}{8} = \frac{3}{8}$ $\frac{3}{8}$ of a gallon

© Frank Schaffer Publications, Inc.   FS-32022 Math Activities

Answer Key

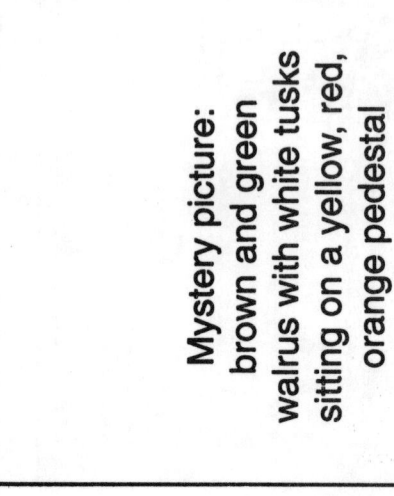

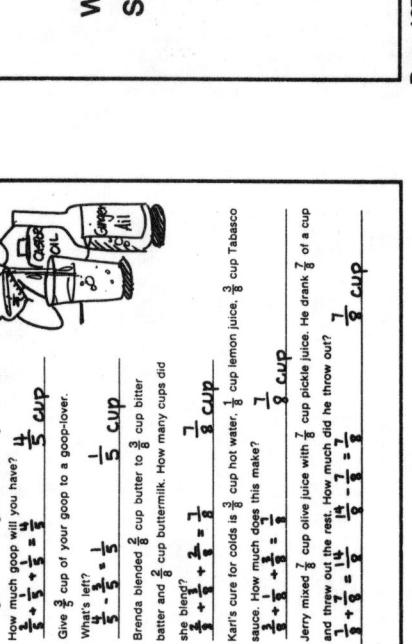

Page 105

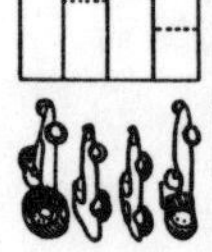

Page 106

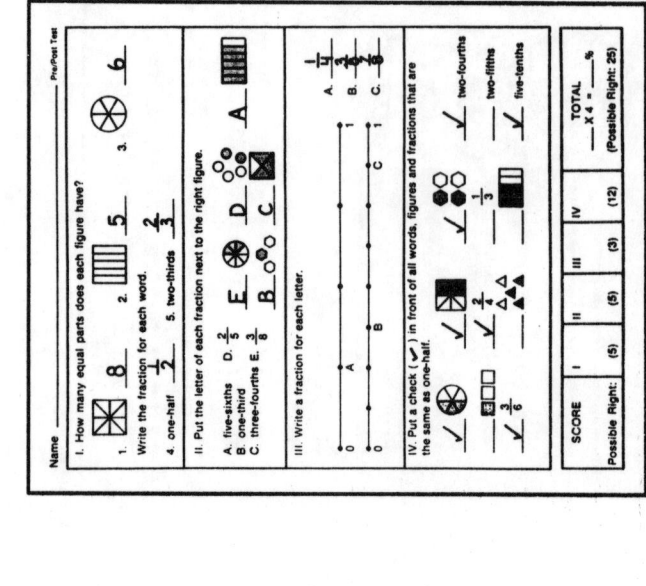

Page 107
Mystery picture: brown and green walrus with white tusks sitting on a yellow, red, orange pedestal

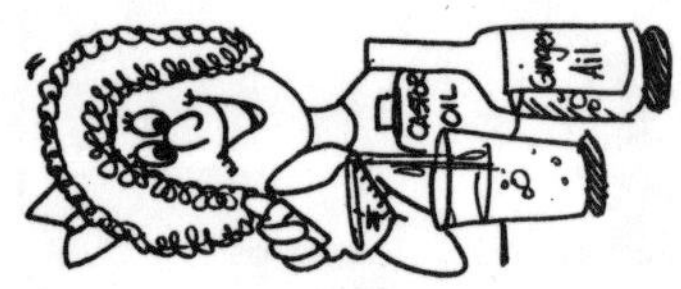

Page 108
Mystery picture: brown pelican with yellow beak sitting on a brown and red post near a red and orange ship

Page 109
Mystery picture: orange shark swimming over an orange squid and red clam

Page 110